THE DISTILLER.

BY HARRISON HALL

White Mule Press
Hayward, CA USA
whitemulepress.com

ISBN 978-0-9910436-8-2

The Distiller, 2nd Edition (Revised).

Illustrations by
Zachary Jones (Tampa, FL)
and Cydney Parmele (Kansas City, MO)

Designed and Edited by
Brad Plummer

THE

DISTILLER.

CONTAINING

1. Full and particular directions for mashing and distilling all kinds of Grain, and imitating Holland Gin and Irish Whiskey.
2. A notice of the different kinds of Stills in use in the United States, and of the Scotch Stills which may be run off 480 times in 24 hours.
3. A Treatise on Fermentation, containing the latest discoveries on the subject.
4. Directions for making Yeast, and preserving it sweet for any length of time.
5. The Rev. Mr. Allison's process of Rectification, with improvements; and mode of imitating French Brandy, &c.
6. Instructions for making all kinds of Cordials, Compound Waters, &c.; also for making Cider, Beer, and various kinds of Wines, &c. &c. &c.

ADAPTED TO THE USE OF FARMERS, AND DISTILLERS.

BY HARRISON HALL.

PHILADELPHIA:
ORIGINAL (1818) PUBLICATION BY THE AUTHOR
133, CHESNUT-STREET.
J. Bioren, Printer

SAN FRANCISCO:
REVISED EDITION (2013)
BY KNOWLEDGE ARTS MEDIA

Anderson's Patent Condensing Tub. 

A *Still to contain 110 Galls. exclusive of the Head.*

B *Pipe 12 In. long to receive the pipe* R *to bring off the head of the Still.*

C *Pipe 26 In. long from the Still head.*

D *Globe 24 In. Diameter.*

E *Chain that rubs the Globe.*

F *Condensing Tub, the height & width as marked in the annexed figure.*

G *Charging Pipe 3 In. wide, with a large cock screwed into the bottom of the Tub, the lower end fitting into Pipe* H *in the breast of the Still.*

I *Pipe to carry off the condensed steam into the Worm fitted into Pipe* B.

K *Square Box with a Square In. hole & 12 In. long to slip up on the Spindle to bring off the Head.*

L *Cogg Wheel 16 In. Diameter.*

M *Pulley 16 In. Diameter.*

N *Strap over the Pulley.*

O *Cogg Wheel 10 In. Diameter.*

P *Head of the Still.*

R *Pipe 10 In. long & 8 In. wide fastened to the Globe.*

S *Pipe 3 In. high and 4 In. wide to receive the Pipe* C *which runs into the Globe.*

T *Joint for the Pipe* I *to slip in.*

W *Pulley 10 In. Diameter to be turned by hand to stir the Still & clean the Globe.*

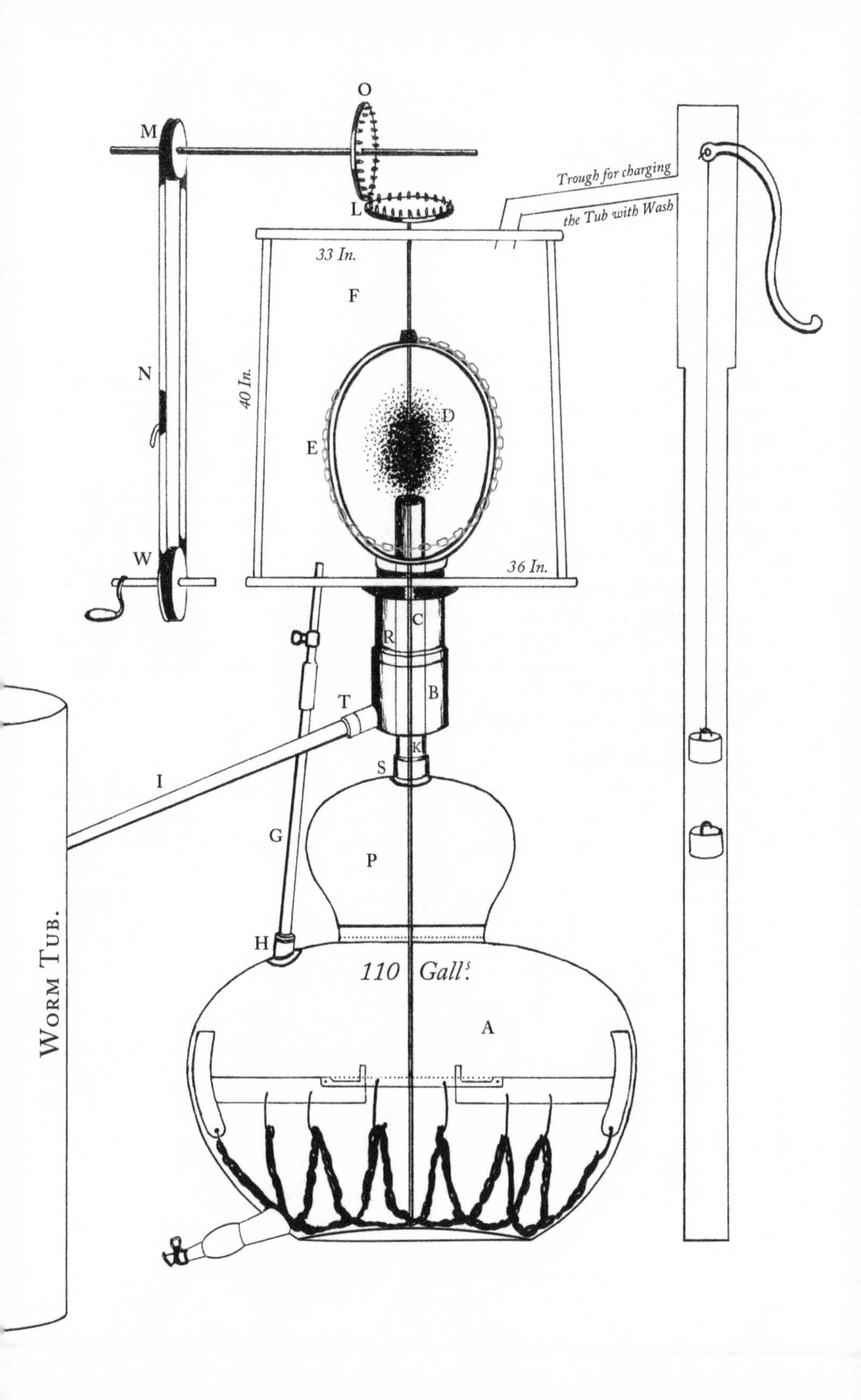

O
M
L
Trough for charging
the Tub with Wash
33 In.
F
N
40 In.
D
E
W
36 In.
C
R
B
T
K
S
I
G
P
H
110 Gall^s.
A
Worm Tub.

RECOMMENDATIONS.

"Practical men, or those engaged in the steady prosecution of any art or manufacture, so seldom favour the public with an account of their processes, that when they do write respecting them, or suggest any improvement therein, the public must gain by their labours.

"I have read Mr. Hall's work on Distilling, and feel no hesitation in saying that I consider it a valuable acquisition to those engaged in that business.

"JAMES MEASE."

"Mr. H. Hall,

"Dear Sir,

"I read your Treatise on Distillation, with great pleasure; it certainly contains more information than any book that I have seen in print. I shall recommend it wherever I have an opportunity of doing so.

"I am, dear Sir, yours,

"ALEX'R. ANDERSON."

"Dear Sir,

"In answer to your note of this morning, I reply, that I read your manuscript upon Distillation with much interest, and derived from it some practical information, which fully repaid me for the time and attention bestowed upon it. I can with sincerity say, that some of your hints have been of use to me.

"I am, sir, with due regard,

"Your obedient servant,

"ROBERT HARE.

"Harrison Hall, Esq."

"Dear Sir,

"I thank you for the perusal of the first part of your Treatise on Distilling. I have seen no work on the subject, in this country, that contains so much practical good sense, or so likely to communicate useful knowledge on the subject.

"I heartily wish you success, and that the Second Part which I have not yet seen, may be as good as the portion you sent me.

"I am, dear Sir,

"Your Obedient Servant,

"THOMAS COOPER.

"Harrison Hall, Esq."

Preface.

It requires but little reflection on the agricultural and commercial state of this country, to shew the propriety of giving more attention to the manufacturing of domestic spirits, than has been hitherto done. Continually liable to interruptions in our trade, with those countries from which we have drawn our supplies of liquors; it would be prudent to become less dependant on them, and more industrious in improving our own capacity to provide for ourselves. These truths have been sensibly felt, since the embargo of 1807, and are every day becoming more important. Hence, the rapid increase in the number of grain distilleries in this country within that period; insomuch that foreign spirits are almost entirely excluded from common use amongst our farmers. Yet is the art of distilling an agreeable and wholesome liquor from the products of our own country, but very imperfectly understood; and yet are we without any certain guide to direct us in our operations.

His own experience and the want of a safe and systematic guide, during several years that the author of the following work was practically engaged in distilling, has induced him to believe that the result of a careful inquiry into the subject might not be unacceptable to such as desire to be informed. As his principal object is to instruct plain, unlettered men, he has laid down his rules, in the most simple manner, and in the fewest possible words; avoiding all irrelevant matter, and doubtful theories which might tend to perplex rather than elucidate. In the course of his inquiries he has had the advantage of visiting other distilleries besides his own; and whilst he has seen the errors of careless, or ignorant pretenders, he hopes he has profited by the liberality and science of enlightened men.

Practical men we know, on all subjects, are too apt to despise books; and we are equally aware that a complete manufacturer can never be formed by reading alone. But though we acknowledge that experience will always be the sure test of truth, it cannot be denied, that books are a convenient medium to bring much valuable truth into view; to compare our opinions with those of others; and above all to bring together the excellencies and errors of the four quarters of the world, that whilst we adopt the one we may avoid the other. To the insulated savage, when he has accidentally discovered that an intoxicating liquor may be extracted from his native maple, it may be allowed to exult in his wonderful knowledge; but he who reads, will rather speak with modesty of his own attainments when he sees how far he has been surpassed by others.

But though we contend for the utility of books in the science of distillation, we can assure those who would rather depend on their own practical results that there is yet great room for the application of their ingenuity and diligence; for the science of fermentation is yet so imperfectly understood that no rules can be given for mashing and making yeast, which may not possibly disappoint an ordinary operator.

On the other hand it may perhaps be asserted with some confidence, that a degree of success equal to his most sanguine expectations may be attained by a careful attention to the directions contained in the following work. They are the result of the author's own experience, or of communications, for the most part personal, with men on whose knowledge and veracity he could alike depend. And however humbly he may indeed be disposed to think of his labours, he cannot but flatter himself that he will be found to have contributed something though but a mite to the general stock of information.

Preface

TO THE SECOND EDITION.

THE uncommonly rapid sale of the first edition of this work, far from inspiring the author with over-weening confidence, has induced him to review his labours with great caution. As was to have been expected, he has received a variety of communications on the subject from several gentlemen. He is not insensible to the favourable opinion that has been expressed by many, as he has endeavoured to evince, in this edition, which will be found to be greatly enlarged, and, it is hoped, improved.

January, 1818.

Preface

TO THE REVISED SECOND EDITION.

GREAT care was taken to present this book in a manner that preserved the flavor and feel of the original work while correcting a handful of obvious errors and standardizing a few inconsistencies. Although corrections to many misspellings were made (proper names, mis-typeset words), colloquial usage and spelling appropriate to the era were maintained. Conversely, unit modifiers were standardized to conform to modern conventions of usage for the sake of readability ("gallons" or "gal" instead of "Gal.", degree symbol [°] instead of "deg.")—no such standards had been employed in the original work, with usage varying from chapter to chapter.

A similarly conservative approach was taken with regard to the book's layout and design, and the reader may therefore note various non-standard layout conventions preserved from the original work.

However, all of the original illustrations were professionally hand-redrawn in order to obtain the highest fidelity of reproduction, and tables that in the original were difficult to decipher were re-set according to modern standards.

Aside from these minor cosmetic and stylistic alterations, no substantive changes were made to the text itself, and it is presented here as closely as possible to how it originally appeared.

Knowledge Arts Media
April, 2013

Index.

Part II

Reference to the Plates.

THE
DISTILLER.

Preliminary Observations.

The time of the invention of brandy or ardent spirits, so useful in the arts, and so important an object of commerce, and which has proved so influential on the health, habits, and happiness of the human race, is involved in obscurity. That the first was made by the Arabians, from wine, and thence called *vinum ustum*; that Arabian physicians first employed it in the composition of medicines, and that so late as the year 1333, the manner of preparing it was still considered by the surgeons as a secret art, appears from the writings of Arnaldus, Raymond, Lully, Theophrastus, and Paracelsus; and it is not without some reason that the invention has been ascribed to Arnold. Alexander Tassoni relates, that the Modonese were the first in Europe, who, in consequence of a superabundant vintage, made considerable quantities of this liquor. The German miners had first acquired a habit of drinking it; and the consumption of, and demand for, this liquor, soon induced the Venetians to participate with the Modonese in this new art, and lucrative branch of commerce. However, it appears that brandy did not come into general use until towards the end of the 15th century, and then it was still called *burnt wine*. The first printed books which made mention of brandy, recommended it as a preservative against most diseases and as a means of prolonging youth and beauty. Similar encomiums have been passed upon tea and coffee; and people became so much habituated to these liquors, that they at last daily drank them merely on account of their being pleasant to the palate.

In the reformation of the archbishopric of Cologne, in the first quarter of the sixteenth century, there is no mention of brandy; although it must certainly have been made there, if it had already been used in Westphalia.

William II. Landgrave of Hesse, about the commencement of the sixteenth century, ordered, that no vendor of brandy should suffer it to be drank in his house, and that no one should be allowed to offer it for sale before the church doors on holidays. In 1524 Philip, Landgrave of Hesse totally prohibitted the sale of burnt wine. But in the middle of the sixteenth century when Baccius wrote *his history of wine*, brandy was every where in Italy sold under the name of *aqua vitis or vitæ*. Under king Erick it was introduced into Sweden.

For a long time this liquor was distilled only from spoiled wine, afterwards it was made from the dregs of beer, wine, &c. and when, instead of these, the distillers employed wheat, rye, and barley, it was considered as a wicked and unpardonable use of grain, it was feared that spirit made from wine would be adulterated with malt spirits; and an idea prevailed that the grains were noxious to cattle, *but especially to swine*; whence originated among men that loathsome disease *the leprosy!*

Expressly for these reasons, *burnt wine* was, in January, 1595, forbidden to be made in the electorate of Saxony, unless only from wine lees, and the dregs of beer.

In 1582 brandy was prohibited at Frankfort on the Main, because the surgeon barbers represented, that it was noxious in the then prevailing disorders. From the same cause the prohibition was renewed in 1605.

The love of brandy or ardent spirits in general, has spread over all parts of the world, and nations, the most uncultivated, and the most ignorant, who can neither reckon nor write, have not only comprehended or devised methods of distilling it, but even had ingenuity to prepare some kind of beverage from the vegetable kingdom of their own country.

"The miserable hordes who wander in the forests of Guayana," says one of the most valuable writers of the present day,* "make as agreeable emulsions from the different palm-tree fruits as the barley water prepared in Europe. The inhabitants of Easter Island, exiled on a mass of arid rocks, without springs, besides the sea water drink the juice of the sugar-cane. The most part of civilized nations draw their drinks from the same plants which constitute the basis of their nourishment, and of which the roots or seeds contain the sugary principle united with the amylaceous substance. Rice in southern and eastern Asia, in Africa the igname root with a few arunis, and in the north of Europe cerealia, furnish fermented liquors. There are few nations who cultivate plants merely with a view to prepare beverages from them. The old continent affords us no instance of vine plantations but to the west of the Indies. In the better days of Greece this cultivation was even confined to the countries situate between the Oxus and Euphrates, to Asia Minor and western Europe. In the rest of the globe nature produces species of wild vitis, but no where else did man endeavour to collect them round him to meliorate them by cultivation."

"But in the new continent," he continues, "we have the example of a people who not only extracted liquors from the amylaceous and sugary substance of the *maize*, the *manioc* and bananas, or from the pulp of several species of *mimosa*, but who cultivated expressly a plant of the family of the *ananas* to convert its juice into a spirituous liquor. This plant differs essentially from the common maguey."

And "A chemist," he adds, "would have some difficulty in preparing the innumerable variety of spirituous, acid, or sugary beverages, which the Indians display a particular address in making, by infusing the grain of maize, in which the sugary matter begins to develop itself by germination. These beverages are generally known by the name of *chica*, have some of them a resemblance to beer, and others to cider. Under the meuastic government of the Incas it was not permitted in Peru to manufacture intoxicating liquors, especially those which are called *vinapu* and *sora*. The Mexican despots were less interested in the public and private morals, and drunkenness was very common among the Indians of the times of the Aztec dynasty. But the Europeans have multiplied the enjoyments of the lower people by the introduction of the sugar-cane. At present in every elevation the Indian has his particular drinks. The plains in the vicinity of the coasts furnish him with spirit from the sugarcane. The *Chica de Mais*

*Baron Humboldt. See his *Political Essay on the Kingdom of New Spain*: translated by John Black, vol. i. p. 518. London, 1811.]

abounds on the declivity of the *Cordilleras*. The central table-land is the country of Mexican vines, the agave plantations, which supply the favourite drink of the natives, the *pulque de maguey*. The Indian in easy circumstances adds to these productions of the American soil a liquor still dearer and rarer, grape brandy, partly furnished by European commerce, and partly distilled in the country.

"Before the arrival of the Europeans, the Mexicans and Peruvians pressed out the juice of the maize stalk to make sugar from it.

"In the valley of Tolucca, the stalk of the maize is squeezed between cylinders, and there is prepared from its fermented juice a spirituous liquor, called *pulque de mahis*, or *tlaolloi*, a liquor which becomes a very important object of commerce."

The various difficulties, restrictions and prohibitions which attended the more general introduction of the distillation of grain, into the different countries of Europe, would afford matter of curious investigation, but does not come within the limits of this work.

In America during the infancy of the settlement, we were under the necessity of importing spirituous liquors of different kinds, as more important cares precluded any attempts at domestic distillation. When it was attempted, it may be presumed not to have been productive of the best spirit. Thence probably arose prejudices against it when manufactured in this country, which retarded improvement in the art, until about the time of the revolution. Cut off from the usual supply of spirits, the price of those which remained, or were occasionally brought into our ports, was so high, as to place them beyond the reach of the great mass of the people. Instead then of following the bright and laudable example which was exhibited by the fair sex, in abstaining from the use of tea, the ingenuity of man was stimulated to obtain a substitute for foreign spirits by the distillation of grain, and, such was the influence of *patriotism*, or rather, the desire of making money, that a single still put up in a shed, with a worm made of gun barrels, was all the apparatus at this time employed in many places in making whiskey. There were however, in some parts of the country well established grain distilleries upon what is called the *old plan*.

Although a number of patents had been granted by our government for improvements in distillation, it does not appear, that any important ones were made until the year 1794, when col. Alexander Anderson obtained a patent for what he termed a steam still, upon which plan he had one boiler which worked two stills. Such were the prejudices in favour of the old way of distilling that very few distillers adopted the improvements of col. Anderson until about the year 1801, when he obtained a patent for his "Condenser for heating wash or any other subject to be distilled." The advantages of this plan being great, it was getting into general use when Mr. Henry Witmer obtained a patent for an improvement upon colonel "Anderson's Condenser." This being more compact, has been gradually taking the place of the other, and is at present esteemed equal, if not superior, to any other in use in the United States.

A great number of patents have been obtained for improvements on stills and in distillation in the United States, a list of which is published in this work, more for the information of the curious, than from any advantage to be derived from it. It is a matter of regret to the editor that he cannot distinguish by a particular description the more important, so as to render them more generally known; such however, is said to be the inefficiency of the patent laws for the security of patentees, that many are unwilling to make their discoveries public in an intelligible form, except where there is an immediate probability of selling a right, so that it is difficult to obtain correct information as to any improvement, for which a patent has been obtained but by a sight of the thing itself.

To this rapid and imperfect sketch of the history of distillation it may be only necessary at this time to add, that a still of 110 gallons with Witmer's improvement upon Anderson, can be run off with ease eight or nine times in twenty-four hours; whereas upon the old plan such a still can be run off but three times in the same number of hours.

The business of the grain distillery may be divided into two parts, first, *the mashing and fermentation*, by which the spirit is formed, though still united with other substances; secondly, *distillation*, by which the spirit is separated from those substances and obtained in a pure state. Though probably not of equal importance, each requires the particular attention of the distiller. It may seem surprising, that *distilling*, the less important of the two, is so well understood, and that our knowledge of mashing and fermentation, is still so imperfect. An examination of the subject, however, will show the former to have been the natural result of the daily observations of distillers, but that the latter depended upon circumstances not within their controul, or observation, and only to be completely effected by an accurate knowledge of the science of chemistry.

The making of a still is a mere mechanical operation, and the shape was formerly left entirely to the whim of the coppersmith who made it in a manner the most advantageous to himself; to wit, of narrow bottom and very deep; as however one differed from another, a little observation only was necessary to show the best. The improvements have been gradual, and we have now attained a point equally distant from the deep stills formerly in use, and the very shallow ones used in Scotland, and once recommended in this country, but which cannot be used advantageously with our thick wash.

However simple the operation of mashing may appear as a mere mixture of grain and water, the mode of doing it constitutes a great difference between distillers, who cannot be successful without a knowledge of the correct process.

The mere practical distiller does his daily work with the use of a certain quantity of water, to a certain quantity of meal, without a proper attention to variation in the heat of the weather, so trifling, as not to be noticed but by reference to a thermometer (which he rarely possesses) but yet sufficient to affect the process, and consequently increase or diminish the product. Of the consequences, he is sensible, but ignorant of the cause, and will say in general terms, that he *hit it* or *missed it*, as the case may be. If such a man, by long practice, should fall into a method of working more profitably than that of his neighbours, his success must be in a great measure the effect of accident, and cannot be called an improvement in the art, because he is unable to make his knowledge useful to others, and even must often fail for want of a certain and invariable rule to direct his operations. But it will be observed that practice makes perfect, and how else, it may be asked, but by practice can you make improvements in any manufacture? It is true, a certain kind of practice is necessary, which in this case may be more properly termed a series of experiments, in conducting which however, the ingredients must be accurately weighed and measured, and the heat of the water and weather precisely ascertained. This, with particular attention to the fermentation, and the results being carefully recorded, will enable the experimenter, not only to make improvements, but to communicate them to the world.

The dependance of this art upon chemistry and the great advantage to be derived from a knowledge of this science are so well illustrated in the introduction to Henry's *Epitome of Chemistry*, that a short extract from his work will not be deemed improper: "but the acquirement of experience, in other words, a talent for the accurate observation of facts, and the habit of arranging facts in the best manner," says this writer, "may be greatly facilitated by the possession of scientific principles. Indeed, it is hardly possible for any one to frame rules for the practice of a chemical art, or to profit by the rules of others, who is unacquainted with the general doctrines of the science. For, in all rules, it is implied, that the promised effect will only take place, when circumstances are precisely the same, as in case under which the rule was formed. To insure the unerring uniformity of result, the substances employed in chemical processes, must be of uniform composition and excellence; or, when it is not possible to obtain them thus unvaried, the artist should be able to judge precisely of the defect or redundancy, that he may proportion his agents according to their qualities. Were chemical knowledge more generally possessed, we should hear less of failures and disappointments in chemical operations; and the artist would commence his proceedings, not, as at present, with distrust and uncertainty, but with a well grounded expectation of success.—In the present imperfect state of his knowledge, the artist is even unable fully to avail himself of those fortunate accidents, by which improvements sometimes occur in his processes; because to the eye of common observation, he may have acted agreeably to established rules, and have varied in circumstances, which he can neither perceive or appreciate. The man of science, in these instances, sees more deeply, and, by availing himself of a minute and accidental difference, contributes at once to the promotion of his own interest, and to the advancement of his art." But it is the union of theory with practice that is now recommended. And, "When theoretical knowledge and practical skill are happily combined in the same person, the intellectual power of man appears in its full perfection, and fits him equally to conduct, with a masterly hand, the details of ordinary business, and to contend successfully with the untried difficulties of new and perplexing situations. In conducting the former, mere experience may be a sufficient guide; but experience and speculation must be combined to prepare us for the latter." Stewart's *Elements of Philosophy of the Human Mind*, chap. IV. sect. vii.

Distillation was for a long time confined to farmers, who only carried on the work during the winter season, and men of small capital, who being obliged to make quick sales, were more attentive to the quantity, than the quality of the spirit distilled. Under the old plan it was supposed that it could not be carried to a sufficient extent to render it an object to a man of large capital, the demand for grain spirit being trifling, owing to the large quantities of New England rum to which preference was given; neither was this business then thought *respectable*.

This was the state of the case at the time of the very ingenious improvement made by Mr. Alexander Anderson, which shed a new light on the subject, and shewed that it could be carried to any extent; the idea too, which at that time was circulated, of making three gallons of spirit from a bushel of grain, when the price of one gallon of gin was nearly equal to that of a bushel of grain; presented the prospect of such a rapid accumulation of wealth, as to cause a very considerable increase of distilleries. For, whatever opinion may be entertained of a manufactory, by those who only view it at a distance, so soon as it is found to be very profitable, it will also be sufficiently *respectable*, for the attention of the most fastidious: Accordingly, we find men of science, men of capital, lawyers, doctors, and merchants abandoning professional pursuits, and the hazardous speculations of the compting house, to learn the art of extracting spirit from grain; and to such men are we indebted for improvements, both in the quantity and quality of the spirit; on the last mentioned particular depends its taking the place of foreign spirits. Let distillers then pay the utmost attention to the flavour and proof, and as these improve, the price will rise, and grain spirits, whether gin or whiskey, will become fashionable, and valued according to their quality.

The very name of whiskey is nauseous to some men; and when they taste some of that which is offered for sale in our cities, the reality is found to be ten times worse than the idea, and they are completely disgusted. Offer to the same persons whiskey which has been *double distilled*,* and carefully attended to in every part of the operation, having also the advantage of twelve months age, they will drink it without being able to say what it is, and may finally prefer it to French brandy.

I have sold many gallons of such as had been made in this manner for the private use of gentlemen, at one dollar per gallon; it has been highly esteemed, and generally preferred to any liquor generally sold under the name of French brandy.

The rapidity of improvements in the western parts of the United States, is a matter of some consideration to the distillers of the Atlantic states. They have already made considerable progress in the art of distillation, and the vast quantities of grain which are produced by their fertile lands, beyond the necessary consumption, cannot be so well disposed of in any way as in pork and whiskey. Hence we already find Tennessee and Kentucky whiskey in our sea ports, and it is generally preferred to that made nearer home; this by the way, is a powerful argument against the common prejudice against using corn, as the western whiskey is chiefly made of that grain; the distillers here, however, without examining into the real quality of this whiskey, are satisfied with attributing this preference to the prejudice generally entertained in favour of things procured with some difficulty. Kentucky hams have also been brought hither and sold at very good prices.

Although the western distillers may not take more pains, or have a more complete knowledge of the art of distillation than others, there are several causes why their whiskey in general is better than ours; setting aside, however, the superior quality of their grain, which is certainly of importance, and some local advantages, it may be merely necessary to observe, that in order to save the expense of transportation and casks, their whiskey is made fourth proof, so that they offer for sale nothing but the pure spirit; whereas our distillers have a vile practice of running feints in their spirits

*It should be observed, that neither age or double distillation, will render good, whiskey originally bad; or that has received an improper flavour during the fermentation.

to reduce them, thereby giving the bad flavour of which complaints have so justly been made. As they depend upon the rise of the rivers to send their whiskey to market, it acquires some age: this also, and the motion of travelling, has considerable effect in improving it. This whiskey has been sold, frequently, from one dollar twenty-five cents, to one dollar fifty cents, per gallon.

The increasing importance of this business is evident, from the quantity of domestic distilled spirits which it appears are made in the United States. Of the twenty-four millions which are annually distilled, it is probable there are twenty millions made from grain: and with the increase of population, and extension of agriculture, distillation will also be augmented.

To the superficial reader of the following pages, and to him who examines but one side of the question, it will appear to be a very profitable business, and he will be surprised that any one engaged in distilling grain should fail of accumulating wealth rapidly. But it may be proper to remark, that although there is no enterprize in which a small capital will yield so large a profit if *well understood*, and properly attended to, in all its parts, neither is there any which will tend more rapidly to ruin the owner than a distillery, conducted without an adequate knowledge of the business, in all its details.

Failures in distilleries are generally attributed to these causes: 1. The situation. 2. The kind of stills, 3. The water. 4. Want of knowledge in the owner to direct, or of a suitable person to conduct, the internal affairs of the establishment. The three first of these causes may be avoided by close attention to the directions in the ensuing pages; and so far as this kind of instruction can be effectual, the author has endeavoured to point out the remedy for the two last.

But there is another difficulty to which many distillers are liable, and which is proper to be mentioned here; this is, the want of a suitable agent to dispose of the gin or whiskey when it is in the market.—For it has come within the knowledge of the author, that a distillery may be perfectly well conducted, and apparently yielding great profits, yet ruined by the ignorance, folly, or dishonesty of the agents: while, on the other hand, there are distilleries from which the liquor and bacon have been highly extolled, and have sold at all times very high, merely from the attention of their agents. Let this point be well attended to then, by any one who is obliged to employ an agent.

On the manufacture of gin, some observations will be found under the proper head. This article, however, at best, is but an imitation of a foreign spirit. We are indebted to a foreign country for the ingredients which imparts the peculiar flavour of that liquor, and even in those we are liable to be greatly deceived. It should therefore become the particular aim of the American distiller to make a spirit purely American, entirely the produce of our own country; and if the pure, unadulterated grain spirit cannot be rendered sufficiently palatable to those tastes, that are vitiated by the use of French brandy or Jamaica rum, let us search our own woods for an article to give it taste sufficiently pleasant for these depraved appetites.

The French sip brandy; the Hollanders swallow gin; the Irish glory in their whiskey; surly John Bull finds "meat and drink" in his porter—and why should not our countrymen have a national beverage?

PART I.

Chapter I.

Concerning the proper situation for a Grain Distillery.

THE first object for the consideration of a man about to enter into the business of a grain distillery, is to procure a proper situation for his works.

This is of greater importance, and requires much more examination and deliberation than is generally supposed, or is usually given to the subject.

The first question that presents itself, is whether a situation contiguous to, or at a distance from a city is to be preferred, and the advantages and disadvantages of each must be considered with reference to the capital to be employed.

In the first case, grain, fuel, labour, and grinding, will be high; rent will also be an object; and it will frequently be difficult to obtain hogs. On the other hand, being at the market, the distiller can take advantage of any sudden rise. He saves storage, commissions, &c. by making his own sales, and may obtain a regular set of customers. Hands too are always to be obtained, and a coppersmith is near in case of accidents.

At a distance from a city, materials will be procured at lower prices, and by attaching a farm to the establishment, a supply of hogs may be raised at little expense. But here, except it be near a turnpike road, or on navigable water, the manufacturer will find it difficult to get his produce to market;—and he will be obliged to use new casks, which however may be made suitable by care in seasoning the staves properly. In case of accidents, he will want a coppersmith, this also may be remedied by keeping a double set of stirrers, and every thing necessary to repair trifling accidents, and there will seldom be wanting an ingenious person to supply the place of the coppersmith.

The most particular enquiries should be made as to the situation of the country with respect to grain and fuel, the quantity and price of each; the number and contiguity of mills; the price of labour; the facility of getting labourers; the habits and manners of the people, as well labourers as others; and the expense and manner of forwarding his produce to market.

The necessary information on these points being obtained, the suitableness of the situation may be determined from the estimate of expenses given in the fourth chapter of this work.

Remote situations will be found very troublesome to the owner of a distillery, unless he gets amongst an active, industrious, trading people. With such he may probably obtain a constant supply of grain and wood, and for this reason, I represent the habits and manners of the people to be an object worthy of consideration.

These observations are more particularly intended for a person who is at perfect liberty to settle any where, but they may be of importance to others, so far as to satisfy them, that it is better not to build a distillery, than to place it in an improper situation.

There are other advantages attending particular situations, which will occur to persons of observation, a detail of which is unnecessary. It will merely be observed, that a place commanding more than one market, is vastly preferable to that which is confined to one: and a neighbourhood where rye whiskey is usually drank, is to be preferred to that which has apple whiskey or peach brandy in abundance.

Chapter II.

On the choice of a proper Site.

Having settled the question which was the subject of the last chapter, the choice of a proper site or seat for a distillery, comes next into view.

Much has been said in favour of creek or river water for fermentation; but so long as pure spring water free from mineral impregnation, can be obtained, in such a way as to be brought into a distillery over head, it will be found most advisable to use it.

To make use of river water, two extra hands are necessary to pump, or a horse must be used with a great quantity of machinery, which is constantly liable to get out of order; besides which, river water is too warm to cool off with in summer, and cold spring, or well water must be had for this purpose; and whenever water is to be pumped either by man or horse, there will generally be found a deficiency in the supply of the coolers.

Spring water therefore being most desirable, look out for a spring of soft water, that will make a good lather with soap, and is clear of minerals, which are sometimes injurious to fermentation. It should have a fall of eight or ten feet within the distance of one hundred and fifty yards, and be in quantity sufficient for daily use. The spring should be enlarged and walled up so as to contain from one to two thousand gallons (according to the quantity wanted) and conveyed under ground in bored logs to the house.

It would also be an advantage if a stream of water could be brought to the distillery sufficient to stir the stills, and pump the wash, or either. This might also be used for the coolers if the spring be not sufficient. The water from the coolers may be used to turn a wheel, to stir the stills.

It is hardly probable that all the requisites here mentioned will be found combined in one spot; the deficiency must then be supplied as well as possible. There are many situations where the machine for raising water, mentioned in this work, will be of great advantage. Montgolfeir's water raiser also may be useful in some places.

The house should if possible be protected from the northern winds, both on account of the fermentation and the situation for the hogs, with a southern exposure: the stills placed in the west end, and sufficiently elevated for the swill to run to the hogs. This will save much trouble, and the pens being placed at the west end, the hogs will be protected from the easterly storms.

Distilleries are generally built too small; as a general rule, four feet each way, for every hogshead, should be allowed for the mashing floor; except in large establishments, where less will do, and six days work will sometimes be on the floor at the same time in winter. From these data the size of the house may be calculated.

For a still with Witmer's improvement upon Anderson, to work eight hogsheads a day, the house should be twenty-four feet in breadth, forty feet in length and fronting the south, with a door near the east end to turn out dirty hogsheads to be washed on a platform raised for the purpose, and one door near the stills to carry in wood, with as little interference as possible with the mashing.

The house should be of stone, one story and a half high, the lower room well plaistered and floored, and the upper, divided for the different kinds of grain, and the accommodation of the workmen.

Chapter III.

Of the Apparatus for a Distillery, and Cost thereof.

For a distillery of the size and upon the plan mentioned in the preceding chapter, it will be necessary to procure:

1 Patent still, 125 gallons and appendages (cost)	$190.00
Patent right to use same,	50.00
1 Doubling still, 100 gallons, and pewter worm,	100.00
1 Boiler, 125 gallons capacity,	100.00
50 Hogsheads,	75.00
Wash pump and singling do.	25.00
2 Kegs and 2 buckets, for sings,	5.00
2 Spirit and 2 water buckets,	4.00
1 Large funnel,	3.00
Iron mashing oar,	2.00
Thermometer and hydrometer,	25.00
Troughs for cooling off and conveying wash,	5.00
	$584.00
Scales and weights, broom, shovel, bags, yeast buckets, covers for hogheads, lamp, besides several small articles too tedious to be enumerated, the want of which will soon be discovered; say for these,	$100.00
110 Hogs to be bought at beginning,	440.00
Nett proceeds in three months,	$1,124.00

Chapter IV.

Of the Profits of a Distillery, and current expenses in different situations.

In estimating the profits of an establishment capable of working twelve bushels a day, we will make our calculation upon the amount necessary to carry on the work for three months, by which time the proceeds will probably come in so as to make further advances unnecessary. Suppose for round numbers 1,000 bushels are worked, to wit:

600 bushels of corn, at 50 cts.	$300.00
300 do. rye, 60	180.00
100 do. malt, 60	60.00
500 pounds juniper berries, ...	100.00
Hops, ...	2.40
100 barrels, ...	100.00
Transportation of 3,000 gallons at 6 cts.	180.00
	$922.40

CR.

By 3,000 gallons, at 55 cts..	1,650.00
Nett proceeds in three months, ...	$727.60

From these estimates it results, that less than $2,500 will be sufficient to furnish and carry on a distillery for one year; and if the quotation of prices be correct, the profits will be nearly three thousand dollars.

This calculation is predicated upon the supposition, that the profit on the hogs will pay the current expenses of manufacturing the grain, which will generally be found to be the case; but taking in these, and reducing the calculation to the greater nicety, we have this statement.

Amount as above,	$922.40
33 cords of wood, say at 2 dollars,	66.00
Grinding 1,000 bushels, at 5 cents,	50.00
Wages 2,500 gallons, at 5 cents,	125.00
	1,163.40
Commissions on sales, leakage, interest on capital, wear of stills, &c. say for even numbers,	139.20
	1,302.60
Now, supposing the produce to be 10 quarts per bushel, we have 2,500 gallons at 55 cents,	1,375.00
Gain upon 110 hogs 84 days, say 546 pounds, at 6 cents per pound,	327.60
	1,702.60
Nett profits in three months,	$400.00
If however the produce were equal to 3 gallons per bushel, as in the other calculation, we have to add,	275.00
Making the whole amount	Dr. $675.00

Which is nearly the same as the preceding.

As the prices of wood and grain vary in different places, the following data are furnished, to enable any person to make the necessary calculations for any situation where the cost of these articles are known.

That 1 cord of good oak wood is sufficient for working 30 bushels of grain.

That 1¼ cords of pine or 14 bushels Virginia stone coal are equal to one cord of oak wood.

That one man may work 10 bushels grain per day.

That wages may be calculated at one dollar per day for the best hands, but for large establishments some of the hands may be hired much lower.

Or a distiller may be hired for 5 cents per gallon on the quantity distilled, or in some places 4 cents.

Grinding is usually 5 cents per bushel, but as there is a small loss, should be calculated at 6 cents.

Every bushel worked will feed 8 or 9 hogs, and the gain of each hog may be fairly set down at six-tenths of a pound a day.

Transportation by tide water may be put down at 1 cent per gallon; distance not material.

Transportation by land varies very much in different places, but will generally be about 6 cents per gallon on any distance between 60 and 100 miles.

The preceding estimates are made for a still upon Mr. Anderson's plan.

It is stated, tint by using steam stills, there is a considerable saving of fuel and labour. The use of the mashing machine will also save labour in a large establishment, as two men can easily attend to the mashing of 60 or 100 bushels a day.

Chapter V.

The different methods, or plans of distilling, and proper form of stills considered.

In the early operations of the grain distillery, much fuel, labour, and time, was employed, in effecting comparatively a small quantity of work. The object of the improvements and machinery adapted by different persons, has been to lessen the expenditures of one or more of these. To the professed distiller, who calculates upon an extensive establishment, every saving, even the most minute, in his daily expenses deserves his attention; but to the man who works a distillery for a few months, or upon a small scale, it becomes a question, whether the original simple plan a little improved, is not better than the expensive machinery proper for large works.

At one time there was no alternative for the distiller, he applied to a coppersmith who gave him such stills as were most profitable to the maker, consequently many distillers have been unsuccessful merely for want of proper stills; the necessity of attention, therefore to this point will be obvious, and the more so, when we consider the variety of patent plans of distilling which have been offered to the public notice of late days, each of which has had advocates who have considered it superior to all others. It would be both tedious and unnecessary to mention all their plans, as it is to be presumed that no person will adopt any one until he first sees it in operation, or is well satisfied of its superiority. It will be sufficient here to notice some of the most usual, with their alledged advantages and disadvantages) and the reasons for my preference of one. However well grounded these reasons may be, in my own opinion, yet as I cannot possibly be acquainted with all plans and situations, I should wish every one to judge for himself, by comparing the expenses of working according to this plan, with such others as he may prefer; he may then decide without difficulty.

The old way of distilling is generally pursued only by such as work upon a small scale, or are unwilling to be at the expense of a patent right. There are however some distillers upon a larger scale, who prefer the old way, because, they say

1. It is less liable to accidents, there being no apparatus as in the patent still.*

2. The spirit is purer, the head of the still being taken off and washed every charge, and the surface of copper acted upon by the steam being less than in the patent still.

*To avoid repetition, and by way of general explanation, it may be proper to observe that by patent still I allude to Mr. Witmer's improvement upon colonel Anderson's condenser, more particularly explained hereafter.

3. A common still of 220 gallons costs no more than a patent one of half the size, and will do nearly the same work in twenty-four hours.

In answer to the first observation it may be remarked, that very little attention is necessary to keep the apparatus in order, and that a common still is much the more likely to *run foul* consequently requires much more attention than the other to prevent the worm from being choked.

To the second, if the head and globe of a patent still are washed every week, which may be easily done the spirit will be as pure as that obtained by any other still; the head of a common still being necessarily much larger than that of the patent, it is probable that the surface of copper acted upon by the steam will be greater in a common still of 220 gallons than in a patent one of half the size; a patent still being almost constantly running is always hot, the condensing power of the charge in the tub prevents any heat from getting to the worm; this even temperature prevents the formation of verdigris; but, the acid of a cold charge in the old way operating upon the hot copper will certainly cause the formation of verdigris, the worm too is considerably heated and as it cools between the different charges verdigris will be formed, hence one of the reasons of the impurity of whiskey.

The third observation is indefinite, but it will be very evident that to work the same number of gallons of wash a greater number of hands will be requisite, and three times the quantity of fuel will be used in the old way, that is necessary for a patent still.

According to this old way of distilling (as originally performed, and even now in many places) the charge is put into the still quite cold, and consequently requires a long time to be made to boil, during which time it is necessary to keep the wash in constant agitation to prevent empyreuma, or adhering to the bottom and sides of the still and burning. This is done by a man with a broom or paddle, and keeps him very busily employed till the wash arrives at the boiling point, when the head is luted on and no further attention is necessary than to regulate the fire and keep up a constant supply of water to the worm tub. The difficulty of ascertaining the proper time for putting on the head is great, and is a matter of much inconvenience and sometimes loss, for if it be done too soon, an empyreuma almost certainly takes place, and if it be delayed until the wash actually boils, it is evident that a portion of the strongest spirit will be lost.

Many expedients have been proposed to remedy this great inconvenience, but without effect, if we except the stirrers adapted to colonel Anderson's patent still.

The great quantity of water necessary for the worm tub, probably first suggested the idea of applying the steam in its passage to the worm to heating a charge previous to its being put into the still. This was done in several ways, amongst which may be mentioned the *hot box*. This was an oblong box, sufficiently large to hold a charge for the still, placed in an elevated situation with a large copper pipe or cylinder passing through the centre, and connected at one end with the arm of the still, and at the other to the worm; a considerable degree of heat might be imparted to the wash by making this cylinder very large. It was soon found however that the thick part of the wash settling on this, formed a coat which confined the heat, and though a scraper was contrived to clean the cylinder, it was still found liable to objections.

These objections were obviated, and other deficiencies in the then method of distilling, remedied by colonel Anderson's "condenser or plan of heating the wash or other subject to be distilled, by means of a half globe."

This half globe made of copper, was placed in the condenser or charging tub fixed above the still. The steam was conveyed by means of a pipe to the half globe where it became condensed by the operation of the charge above, to which it imparted considerable heat, and the liquid thus condensed, was conveyed by means of another pipe to the worm. Stirrers were contrived to keep the wash constantly in motion in the still, and also in the condenser; these stirrers were moved, by means of machinery, by water, by a horse or any other power; a charging cock and pipe between the tub and still, prevented the necessity of moving the head to charge the still. The advantage gained by heating the wash in this manner was such that a still which in the old way could be run off only three times in twenty-four hours, was now run off eight or nine times.

Many absurd objections were made to this plan of distilling by those who were unwilling to leave the beaten track, and such as were desirous of depriving the inventor of the credit due to his ingenuity.* But the real errors were not discovered until after a few years experience, when they were remedied by Mr. Witmer's improvement, a draught of which is given in this work. They consisted in the *shape* of the half globe, and in the steam-pipe, it being very difficult to clean the former, and the latter being much exposed to the action of the cold air, much of the vapour was condensed in its passage to the half globe, and returned into the still, thereby retarding the process and weakening the product.

A reference to the plate will afford a particular explanation of Mr. Witmer's improvement, it may be therefore sufficient to observe, that the whole apparatus is very compact, not liable to get out of order, and may be taken apart and put together again in a few minutes whenever it requires cleaning, and so well contrived that this may be effected with great ease. The cold wash which is pumped into the condenser or charging tub, operates as a condenser to the vapour arising within the globe, and at the same time acquires so much heat from the steam, that it will boil in a few minutes after it is let into the still.

That the superiority of this plan may be still more apparent, I subjoin a short notice of the Scotch stills, with a comparative estimate of the expense attending them and Anderson's stills upon a large scale, from which it is evident that the Scotch stills will not suit us; our object is to do the greatest quantity of work in a given time with the least expense of labour and fuel. The still for this purpose then (for Anderson's and Witmer's improvements may be added to a still of any shape) should be of such dimensions that the charge should be about 16 inches deep, in a still calculated to run 100 gallons, and proportionably increased with the capacity so as not to exceed 30 or 36 inches in a still of 500 gallons.

*Without entering into the question as to the original application of steam as the means of conveying heat, we are willing to give full credit to the colonel for his mode of applying it. We notice it as a new era in distillation and the importance of it is fully evidenced by its use, as well as by the numerous evasions of his right, which we constantly see in the different methods of using steam.

And, that the evaporation be not retarded by the collection and pressure of the vapour in the vacant part of the still above the charge, and also to lessen the possibility of the still *running foul*; this vacant space should be very large—that is, a still to run one hogshead or one hundred gallons should hold one hundred and twenty-five gallons without the head, and in the same proportion for larger stills.

From the experiments which I have made, I have no doubt, a still of this kind may be constructed to run off twelve times in twenty-four hours.

It is difficult to fix the point of perfection in any art, the *ne plus ultra* beyond which improvements cannot be made. Here, however seems to be a resting place. And I would advise the distiller who becomes possessed of a still of any size,* which may be run off twelve times in twenty-four hours with the attention of one man only at a time, to charge, discharge, &c. to consider well, before he adopts any other plan termed an *improvement*.

It is the opinion of colonel Anderson, and as such, well worthy of attention, that the "*quality* of the spirit is determined in the *act of fermentation*; the form of the still having nothing to do therewith; the act of distillation being a mere separation of the spirit and water.†"

My own observations have satisfied me that much of the quality of the spirit is determined during the fermentation; but I have also been led to believe, that a fiery or mild spirit will be obtained in proportion as the distillation (in the doubling still) is rapidly or slowly conducted.

The form of this still perhaps is not material, though that of the wash still, to which the preceding observations are more particularly applicable, may possibly be found to have some effect upon the quality of the spirit. But it fortunately happens, should this be the case, that the same still which is preferred for other reasons is also found preferable for this.

Mr. Nicholson in his journal No. 108, says: that deep stills are best for distilling those simple or spirituous waters, where a full impregnation with the peculiar flavour of the vegetable substance employed, is desirable‡; yet a shallow still is preferable, where the object is, to prevent as much as possible the peculiar flavour of the liquor distilled from rising as in distilling from grain or molasses, and this not only on account of the saving in time and fuel, but of the *superiority of the liquor, in point of flavour*.

The opinions of Mr. Nicholson are certainly entitled to consideration; it seems however not a little surprising that he should have formed any conclusion upon the assertions of Mr. Curaudeau, whose experiments are published as the ground of the opinion here expressed.§

For instance, Mr. C. says "I satisfied myself, that, in the common still the evaporation of the spirit does not begin to be very copious, till the heat is from 190 to

*Stills upon this plan, are run off in the same time whether of 100 or 500 gallons.
†See *Archives of Useful Knowledge*, Vol. 1 No. 1, by Dr. Mease.
‡Or, in making peach brandy... Ed.
§See *Archives of Useful Knowledge*, Vol. 1, No. 1. Or, for original, *Joninis Bibliotheque Physico, Economique*, Paris 1808, tome 1, p. 106.

200 degrees, F while on the contrary, in the shallow still it is very abundant from 133 to 156 degrees, F."

Now in what way Mr. C. "*satisfied himself*" we know not, but his bare assertion does not satisfy us that wine can be made to boil at 133 or even 156° F when we know that the heat necessary to boil alkohol is 165° F and to boil water, 212°. Now wine being a mixture of the two, and the boiling point, depending upon the proportions of the two liquids, must be at some intermediate degree of heat, consequently cannot be below 165 degrees.

Hence we conclude, that he must be *mistaken* in this part of his experiment.

Nor can we assent to his assertion that evaporation takes place at a much *lower degree* of heat in a shallow than a deep still; seeing that one particular degree of heat is necessary; this indeed may be given *sooner* in a shallow than deep still. When however the evaporation has once commenced, it may be increased or diminished by the addition or diminution of fire, but the temperature of the fluid will undergo *very little* variation.

The subject is worthy of further investigation.

A Mr. Kraft of Bristol, Pennsylvania, some years ago obtained a patent for a still, which he stays he works off in fifty minutes. This still is so similar to, if not copied from, the Scotch stills, as to be liable to the same objections; yet in a book on distillation, which Mr. K. published some years ago for the purpose of recommending his stills to notice, he says there were then (in 1804) 217 distilleries at work on his plan.

Most of the objections that have been alleged against Anderson's condenser are mentioned by Mr. Kraft, and experience has proved them all to be unfounded; his prophecy that "it will dwindle into disrepute by fair experiment," remains yet to be fulfilled; and the absurdity of the opinion, that vapour or steam cannot be retained in wooden vessels, is fully proved, by the lately invented wooden stills by Mr. Phares Bernard, of Whitestown, Oneida county, New York; or, those of Mr. P. M. Hackley, a drawing of which is in this work. Proof on this subject however, was unnecessary to any one who knows that iron bound hogsheads have been *burst* by the fermentation of cider rather than suffer the fixed air to escape through the pores of the wood. The opinion that any evaporation takes place from the condenser may be fully disproved by the thermometer, as it will be found that the heat of wash never exceeds 200 degrees. Cider however will boil in the condenser, and it may be, from this that Mr. Kraft and many others have adopted their erroneous opinions without knowing that a higher degree of heat is necessary to boil the one than the other.

Note.

About the year 1788, a considerable duty having been laid upon the capacity of the stills in Scotland, the ingenuity of the distillers was excited to lessen the duties, and in a short time they so far improved their method as to run off a still twenty times in twenty-four hours. They attained to this degree of dispatch by greatly reducing the size of their stills and enlarging their furnaces.

The duty then not costing more than one penny per gallon, no farther improvement was thought necessary until about the year 1797, when the duty was increased to the enormous sum of 54 pounds per gallon on the capacity of the still. Every expedient was now tried to accelerate the process, and from repeated experiments, they found that the more shallow the stills are made, and the bottoms enlarged, the more they could increase the size of the furnace, and apply a greater quantity of fuel, and consequently bring the wash in the still to boil in a shorter time.

The liquor in the still being likewise on a more extended surface, the evaporation takes place in a more expeditious manner, thus they were enabled to run off their stills seventy-two times in twenty-four hours. A degree of dispatch, a few years before, thought to be impracticable.

But the still now in use is so powerful, that hardly any evidence short of that with which the fact is supported, could make it credible. The evidence however is complete; it is that of the distillers themselves, whose interest it evidently was to depreciate rather than overrate the power of their stills.

"The depth of the body is only two and a half inches at the centre, and at the sides the sole and shoulder meet at an acute angle. No sooner was this still set to work than it was evident that the principle on which the shoulder was constructed was just; for though the body and head held only 52 or 53 gallons, the still could work with 22 gallons of wash if the workmen were careful; but steadily, and without foul running for a day together, with 20 gallons; and the time between charge and charge was only three minutes at an average, equal to 480 times in twenty-four hours."

Now 480 charges of 20 gallons each, will amount to 9,600 gallons of wash run daily; to do which will require four hands constantly attending the still, who on account of the severity of the duty, must be relieved at least every eight hours: but probably oftener. Hence it will require twelve men at 5 shillings per day, ... $8.00

To prepare the wash and do the other work about the house will require ten men at 5 shillings per day, ...6.67

$14.67

That is, the charge for manual labour alone, on 9,600 gallons wash, is $14.67, with the Scotch stills.

The stills of Anderson and Hall at Lamberton, upon colonel Anderson's patent plan, were of capacity sufficient to run 500 gallons at every charge, and eight charges in twenty-four hours. That is, 4,000 gallons daily.

One hand only was necessary to attend the still, who was relieved every twelve hours, that is two hands a day, and five hands to prepare the wash and do the other work of the house; being in the whole seven hands, at 5 shillings, ... $4.67

That is, the charge for manual labour alone on 4,000 gallons wash is $4.67 by A. and H's. stills.

Then by a simple statement:

If 4,000 gallons cost $4.67 : : 9,600 gallons will cost $11.21
But by the Scotch plan, 9,600 cost $14.67

The balance in favour of A. and H's. stills then is $3.46
upon the daily work of 9,600 gallons for manual labour alone.

This calculation is imperfect, as the wash used in Scotland is different from ours. But it being almost impossible to work our wash in flat stills, adds to the argument against the Scotch stills.

We have no data by which to ascertain the quantity of fuel used; but a moment's reflection must satisfy us that as the Scotch still is *all bottom*, it requires more fire than ours, if we are allowed to *guess*, we should say four times as much.

Chapter VI.

Distillation by Steam.

This mode of distilling is becoming very prevalent throughout the United States; it is strenuously advocated by those who are interested in its adoption, and has some powerful friends amongst scientific men who have examined the subject.

The formation of a new establishment upon this plan is said to be considerably less expensive than upon any other, and in the daily expenditure for carrying on the work there is also a material saving.

These considerations render the matter well worthy of attention, and the author has been at no little pains to obtain such information as would enable him to state precisely the comparative advantages and disadvantages of this mode of distilling; for this purpose, he has visited several distilleries in this and the adjoining states, and he has corresponded with the owners of others, but he has not yet succeeded to his entire satisfaction. Few distillers in this country are sufficiently attentive to the minutiæ of their business, and still fewer are willing to communicate information to others, generally supposing that any discovery made by them is unknown to all the world beside.

Whenever his own observation, or knowledge of a fact, enables the author to give a decided opinion, he is perfectly willing to do so; but, neither is he willing to hazard this opinion upon uncertain grounds, nor would his *ipse dixit*, unaccompanied by evidence, be entitled to regard. Let his readers then judge for themselves, upon the information which shall be laid before them.

Of the articles which follow, some may be found to contain matter unimportant to the mere American distiller, and the same remarks may be repeated more than once. It is not intended that the contents of this work be confined to what is only absolutely necessary for a distiller to know, but from the exhibition of the different modes, both of American and European distillers, the superiority of the former will certainly appear, and the very defects of the latter will suggest hints, from which our ingenuity and industry will not fail to improve. Undervalued as we are by their writers, let us still assert our equality at least with them, and when we compare our attainments in the useful arts of life, with those of any other nation, it is not too much to say, we may place confidence in ourselves.

Such reasons as these, it will be observed, could alone induce us to give an account of the expensive and complicated machinery of Monsieur Adam, but nauseated as we are with the flattery of French philosophers upon the improvements of each other, and the importance in which all their writings is given to the most trifling subjects, we would willingly omit any notice of Isaac Berard's plan of distilling, had not Senator Chaptal, whose opinion is always entitled to respect, pronounced it to be "the *ne plus ultra* of perfection." It is not however a distillation by steam, nor can it be used in distilling from grain; for a particular description, we refer to the 20th vol. *New Series of the Repertory of Arts*. It will be sufficient here to mention, that his improvement consists in three cylindrical tubes joined together like the three sides of a square, and placed in a condenser filled with water; these cylinders are divided into twelve different apartments, with small holes to communicate from one to another. As the vapours from the still strike these different surfaces, the grosser particles become condensed and return to the still, while the finer portions containing the spirit pass over into the worm. M. Berard was induced to adopt this plan, from considering a principle long known in chemistry:—"That all liquors do not enter into ebullition at the same degree of heat, and that the most volatile boil at the least degree of heat. From an inverse mode of reasoning it follows, that when many liquors of different specific gravities are turned into vapour by the application of heat, and pass together into an atmosphere of a less elevated or colder temperature, the most volatile will be the last condensed."

A comparison of these different processes by Monsieur Lenormand may be found in the 21st vol. of the *Repertory of Arts*. There is nothing worth transcribing, except an observation of Senator Chaptal's, "that Adam's process affords the *inappreciable* advantage of extracting a great quantity of brandy from a given quantity of wine, which advantage is caused, without doubt, by the *greater degree of pressure and heat* which the wine undergoes, especially in the still and the first oval vessels." Mons. L. adds, "that this pressure is *necessary* to obtain the great effect which this apparatus produces."

If this be *true*, and we should like to see the fact established by a *practical* distiller, Bernard's or Hackley's plan possess the same advantage; but, though we admit that the great pressure may cause spirits to be produced of greater strength, we cannot believe, without further evidence, that the quantity will be increased.

So much for distillation by steam in France. We come now to the introduction of the American and English plans, and we think it will be abundantly evident, that America has the honour of the original invention, and of the greatest perfection in the mode of distilling by steam.

In the *Emporium of Arts*, vol. 5, No. 3, for October 1814, Professor Cooper asserts, that he was the first person who distilled by steam in this country, but that the original invention was Count Rumford's: he says, "the method of distilling by steam introduced into the wash, is doubtless the best yet found out. Count Rumford first shewed how several vessels of water might be boiled by steam conveyed from one boiler common to them all. He is the inventor of the process. Whether steam be applied to boil water, to boil the liquor in a dyer's copper or the wash in a distiller's hogshead, the principle is the same."

"This is the reason why I did not think myself entitled to a patent for being the first in this country: who distilled by steam. In the fall of 1809, I adapted a lid to a boiler in Dr. Priestley's laboratory, and soldered it on. A safety valve was made of an inch tube of copper soldered in the lid. The liquor was supplied by a small wooden cistern above, with a pipe going near to the bottom of the boiler. A tube from the boiler at right angles, conveyed the steam into the vessel containing the wash or beer; the tube reached within four inches of the bottom of the vessel. It had a cock adapted to it, so that the steam could be stopped off at pleasure. M. Schmidt, who afterwards conducted Mr. Joseph Priestley's distillery, assisted me. John Hall, Esq. late marshal of Pennsylvania, himself well conversant in the business, and Enoch Smith, Esq. of Sunbury, dining with me one day while this experiment was going on, I shewed them the process itself, as they will testify.

"I find a patent has been taken out for this method by some one who has just as much title to the invention as any reader of this article. The right of taking out patents is abused so egregiously, that it has become a perfect nuisance."

Col. A. Anderson also claims the invention of a mode of distilling by steam. In a communication received from him in reply to a number of queries on the subject, he observes:

"In the year 1790, I discovered that steam arising from water, or water itself might be made hotter than 212° (the boiling point) by confining it so that the vapour could not fly off, and that this principle might be applied to evaporating liquors that were liable to be injured by the naked fire; I immediately proceeded to put this into practice, and after a number of experiments in the year 1794, succeeded in completing a plan, agreeably to the accompanying plate, which fully answered my expectations, and for which I took out a patent, and made use of it on an extensive scale near Columbia, Pa.

"In using the above mode of distillation, I discovered that the steam in passing under the wash, communicated a heat of 190 to 200°, without pressure, but no higher, that the heat remained stationary until a pressure was put upon it. The pressure necessary was from 3 to 5 pounds to the square inch; with that pressure the distillation was rapid and very complete. Directly after this was put in practice, several attempts were made to evade the patent, amongst others, one from Kentucky, in the year 1796, *by forcing the steam through the wash*, instead of letting it pass under a plate of copper, placed between the wash and the steam. This was thought to be different from my patent, but the principle, as far as I can judge, is certainly the same, as it requires a pressure of 5 to 7 pounds to the square inch before the steam could be driven through the wash so as to communicate heat sufficient to make it boil. This mode of distillation has lately been made use of in France, and is spoken of very favourably, but I am confident it is not equal to the mode adopted by me in 1790.

"The observation of the principle above stated, that steam without pressure would not, however long it continued to act, communicate to any liquor mixed with grain, a greater heat than 200°, and that wash being heavier than water, would not boil until raised to 214°, suggested to me a much more advantageous plan of distillation—to apply the steam arising in distillation to heat the wash or subject to be distilled. By this plan, double the work could be done, with the same expense of time and fuel, than by any other mode then in use. In 1801 I took out a patent, and this mode of distilling is in general use throughout the United States.

"I have made several improvements upon the original invention, and feel no hesitation in asserting that the present mode adopted by me is superior to any other in the use in the United States, as to the economy of labour and fuel; one cord of pine wood affords sufficient fuel for 45 bushels of grain, and three men will work 45 bushels per day."

This is not properly a distillation by steam, but yet we conceive it to be the origin of steam distilling; it was long antecedent to professor Cooper's experiment, and previous to the publication of Count Rumford's fifteenth essay, in which he speaks of the use of steam as a vehicle for conveying heat. This essay contains so much valuable information, that we have thought proper to insert it in this work.

The first patent which we have been able to find for steam stills, was obtained in London in 1802, by Mr. Charles Wyatt. His being the first publication which states with any clearness the alleged advantages resulting from this mode of distillation, they are given in his own words, as "facts ascertained by repeated and extensive experiments."

"The principal advantages are,

1st. An improvement in the quality of the spirit.

2d. A facility and security in conducting the operation.

3d. A reduction in the labour, and in the duration, of the process.

4th. A reduction in the expense of fuel.

5th. A reduction in the original expense and repair of utensils."

1. *An improvement in the quality of the spirit.*

A committee of the House of Commons, which sat in the year 1799, to enquire into the state of the Scotch distilleries, reports, that the disagreeable and unwholesome flavour frequently discovered in spirituous liquors arises, from *rapid distillation. The essential oils and other particles of an offensive smell and taste, rise in rapid distillation with the spirit and communicate to it their peculiar flavour*. In confirmation of this fact some extracts from the report are annexed to these observations.

On considering the nature of distillation, it will be obvious, that, to obtain a pure and genuine spirit no more heat should be applied than will detach the spirit from its basis, although, for the purposes of commerce, great niceties cannot be observed. But, rapid distillation requires great heat. Great heat expels the essential oils, and other adventitious substances; burns the extractive matter that falls to the bottom of the still; and is thus the true source of the depravation of the spirit.

By using steam as the vehicle of heat it is proposed to remedy these inconveniences. The heat as applied in this apparatus, can never exceed that of boiling water; the liquor will be constantly attenuated by the accession of condensed steam; the extractive matter cannot be deposited nor burnt; the essential oils cannot in any undue proportion be expelled; and therefore the spirit will rise in a milder and purer state than by the immediate application of fire.

2. *A facility and security in conducting the operation.*

The fire not being in contact with any of the distilling vessels; but with the boiler only, which by a very simple contrivance, supplies itself with water during the operation, no danger is to be apprehended, from any sudden increase of intensity in the fire, nor can it injure either the quality or ultimate quantity of the spirit. One particular convenience is, that the fire may be situated in a distant or external part of the building.

3. *A reduction in the labour and the duration of the process.*

To make this intelligible, it is necessary to observe that the stills are double, i.e. that one still is placed upon the other, and that the steam is let into the lower, but not into the upper still. Both stills are to be charged at once. The roof of the lower still is so constructed, that any particle of steam, on being condensed against it, runs immediately into the refrigeratory; and the heat that involves in that condensation, passing into the superior charge, prepares it for distillation.

The lower charge being let off, the upper charge supplies its place, and begins almost immediately to run.

If the upper still contain low wines, the operation will be partly simultaneous. Thus, four circumstances concur to accelerate the operation.

1st. No significant time is lost in raising the liquor, the first charge excepted, to the boiling point.

2nd. None of the steam that is raised within the lower still, returns in a liquid state into the general mass to be again raised.

3rd. Whatever is obtained from the upper still during the simultaneous operation, is so much time of the common process curtailed. And,

Lastly, as no cake will form within, all the time usually required to cool and clean the stills, is here dispensed with. In fifty successive distillations, the abridgement of the process cannot be so little as 150 hours.

4. *A reduction in the expense of fuel.*

This is a point of considerable importance. Three hundred pounds weight of New Castle coal, will produce from 550 gallons of molasses wash, undergoing two distillations, one puncheon of rum of 110 gallons, hydrometer proof. According to some reports from Jamaica (possibly not accurate) this is only one-fifth part of the quantity of coal consumed there to produce the like quantity of spirit.* The difference however, even if less in favour of the new system, cannot fail to be an object of solicitude to every distiller. In Scotland, where the distillation is urged by every practicable contrivance, the waste of fuel is enormous.

*Where wood is used as fuel it is necessary to allow 1,089 lbs. of dry oak to 600 lbs New Castle coal, as equal heats are produced by the respective quantities. *Vide: Thompson's Chemistry.*

5. *A reduction in the original expense and repair of utensils.*

It would be improper here, to state the cost of the apparatus, but it would amount to less money than is usually given for utensils of the common construction; and certainly, as the stills do not come in contact with the fire, as no harsh or destructive incrustation whatever can attach itself internally to the metal, the decay and consequently the repairs of the vessels, can hardly be a subject of calculation. It is probable that little or no expense may be incurred on that point, within twenty or thirty years, except for the boiler, which cannot be more than a septennial charge.

In the distillery in which the experiments leading to these conclusions were made, the fermenting vats, the reservoir containing the condensers, the cisterns for receiving the spirit, are all constructed of brick, laid in and faced with Roman cement; and although upwards of 14,000 gallons of wash have been distilled, no perceptible injury has been communicated to any of them, nor have the stills ever been cleaned, except being washed out with warm water. In vats built of these materials, all liquors may be kept cooler than they can be in wood, and their durability is much greater.

Extracts from the Report of the Committee of the House of Commons on the Scotch Distilleries *referred to above.*

"From experience and observation, I have found that spirits are more disagreeable and fiery when distilled with a great degree of heat, than when distilled by a more moderate one."

Mr. Gordon, of Xeres.

"They who assert that rapid distillation has no influence upon the taste or flavour of the spirit, either try to deceive, or are ignorant of the art of distilling. 'Why do we distil fine and delicate liquors in balneo mariæ?' Because the heat is equal and uniform, and cannot be increased by the vivacity of the fire."

Mr. E. G. I. Crookens.

"I am of opinion, that the essential or flavouring oils, cannot be separated so well from the spirits of fermented liquors or washes of any kind, by rapid distillation, as by slow distillation."

Dr. Joseph Black.

"The more rapidly the distillation is carried on, the more the spirits obtained by it will be affected with essential oils, and other particles of an offensive smell and taste, as well as with water or phlegm."

Dr. Inghenhauz.

"I have no hesitation in asserting, that rapid distillation brings over a very strong deleterious spirit, containing empyreumatic oil, which is highly obnoxious to those who drink it."

Mr. William Bannerman.

"I have no doubt that spirits are more unwholesome if distilled very rapidly, than when distilled slowly and with a gentle heat."

Dr. Skene.

In 1803, a patent was obtained by Samuel Brown and others of the state of Kentucky, which, differs very little if any thing from the preceding; and in 1810, Mr. Phares Bernard, of New York, obtained a patent for steam stills; these were the first to be generally introduced into common use in this section of the United States. A number of certificates, from highly respectable gentlemen in favour of this mode of distilling, have been published in a small pamphlet. It is stated to effect a "saving of about one third of the labour and nearly one half of the fuel, and cause an *increase* in the produce of *more than one quart from a bushel* over the productions from common stills, the spirit of a superior quality." When apples are distilled in the pummice, more is saved than would pay all the expenses of manufacturing and distilling it in the usual way.

The apparatus consists of a boiler (size not mentioned) and three tubs, one of 600 gallons, one of 160, and one of 60; the first contains the wash, and the two others singlings.

Another patent was obtained in the year following, by Philo M. Hackley, of Herkimer, New York, for a perpetual steam still and water boiler, which he states contains the only American improvements on steam distilling. The improvements are,

1. In the boiler, which is so constructed that the water is divided into thin sheets, and the fire passes on each side of each sheet of water, making a furnace within a *furnace*, or a furnace and boiler united.

2. By passing the fire through a sheet iron cylinder, placed in a tub of water, by which means the water is heated for the purpose of mashing, or supplying the steam boiler.

3. A water gauge, so contrived as to keep up a regular supply of boiling water from the above mentioned tub to the steam boiler.

4. A double set of still tubs, one of which may be charged whilst the other is running off—the spare steam to be turned into them by means of stop cocks, and thus a perpetual distillation may be kept up.

5. In a worm heater tub, by which the water or wash may be heated.

6. Safety valves in the head of the boiler, to prevent any accident from too great a pressure of steam.

In a letter from Mr. Hackley, he says, "I use a boiler in my small distillery that holds and is taxed 40 gallons, with which we run 24 bushels per day with one half of a cord of good wood; we can run 40 bushels in twenty-four hours with three-fourths of a cord of wood. The number of tubs may be from two to five, in proportion to the magnitude of the distillery. In mine, with but two tubs I can run one gallon of fourth proof spirit per minute. Two hands do all the work and take care of the stock; the apparatus separate from the building cost $520.00."

"To run six bushels at a charge in two tubs, the beer tub should be six feet six inches at bottom, four feet six inches high, and four feet one inch at top, all inside measure. The doubler should hold about two hundred gallons. The liquid to be from one-and-a-half to two-and-a-half feet deep in each.

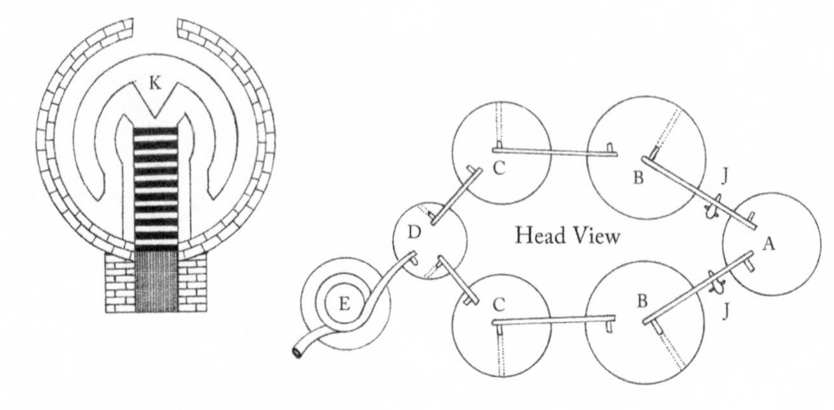

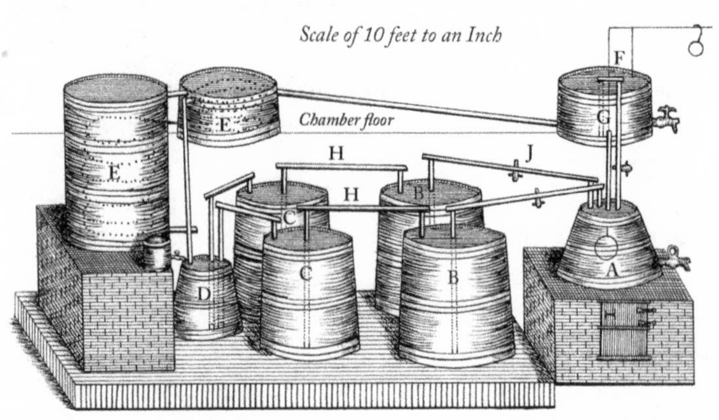

A. *Water Boiler*
B. *Large Still Tubs.*
C. *Small Still Tubs.*
D. *Doubler.*
E. *Condensing Tubs & Receiver*
F. *Water for Mashing.*
G. *Supplier of Water.*
H. *Conductors of Steam.*
J. *Stop Cock in Conductors.*
K. *Bottom view of Boiler.*

P.M. Hackley's Patent.

"When there are more tubs, they may of course be smaller, to do the same quantity of work. A complete view of the apparatus may be seen in the annexed plate."

Another competitor for public favour, is Mr. Robert Gillespie, who obtained a patent for what he terms the "Columbian Independent Log-Still." The great originality of this plan consists in the materials of which this still is composed, being cut out of a solid log, hence it may very easily and cheaply be obtained by persons residing in a part of the country where large logs are attainable, and to such this plan will be a valuable acquisition.

Certificates have been published by Mr. G. from men of great respectability in Tennessee and Kentucky which go to prove them superior to the copper stills.

These certificates, with the addition of much information respecting the mode of making these stills are contained in a small pamphlet published by Mr. G. in which he states his plan to possess the following advantages:

First—It does not cost over one-third of what stills usually cost to do the same business.

Second—One fire does full as much as three usually do, as the cylinder or cylinders receive the immediate action of the fire and the reverberating action of the arch or cover of the furnace, and the steam of one vessel boils four and also heats the beer, &c. in the charger to two thirds of a boiling heat.

Third—It must be evidently more durable than copper stills, as a cast iron cylinder half an inch thick will boil water (all but forever), and boiling in wood neither cracks nor rots it, as all brewers can testify, as they generally use wooden curbs round the mouth of their boilers.

Fourth—It does not require more than half the labour usually necessary, such as uncapping, stirring, luting, checking and drawing of fires. The running of the still (on my last improvement) is also regulated instantly by turning a screw which elevates or lowers a safety valve. It is also easier filled than any other still which uses a heater or charger.

Fifth—It completely prevents the greatest of all evils in distilling, namely, making what is generally called burnt liquor.

These are its decided advantages over all other stills which receive the immediate action of the fire and it is superior to the tub stills in the following particulars:

First—In point of substance and strength.

Second—In security and convenience.

Third—The steam passages are shorter, and the steam therefore coming in contact with less surface, is less condensed; and consequently less water is boiled away. The passages are also stronger and tighter than any other conveyances yet used, and less subject to get out of order.

Fourth—As the vessels in the log are dug out of the solid timber, they can be made of any form found most advantageous, in order to boil all the thickest of the beer, pummice, &c. this shape must be less or more concave, as any one the least acquainted with the properties of heat, know that its action is perpendicular, and not horizontal, therefore one steam conductor can never act on the thick beer which has settled on the bottom of a tub four or five feet in diameter, so as to raise it by boiling it, and thereby extract its strength.

Fifth—It has long been ascertained by practical distillers, that small stills yield more spirit from their proportion of the materials distilling, than large ones do.

The chemical change produced in beer, when kept any length of time in a heated state, producing acid alone, instead of spirit, easily accounts for the difference and therefore the log still running off every hour and a half or two hours, is an advantage instead of a disadvantage.

Sixth—The water vessel or steam generator is easier filled on my plan than in any mode heretofore used. An inch auger hole passes between the steam chamber and condensing cistern, and a valve in the inside is kept shut by the pressure of the steam when the vessels are boiling, and as soon as the resistance is removed by the distilling vessels being discharged, then the valve opens and admits a sufficiency of water to make the next run—or the charging hole may be stopped with a pin and opened at the end of every charge. This hole ought to be five or six inches under the lower side of the cap or cover of the vessel.

The following table gives the proportion of log stills, to make from 20 to 60 gallons in twenty-four hours, with moderate attention, and is made from actual experiments on four different scales.

Proportions of log stills.

No. of cylinders in a furnace	1	1	2	2
Length in feet	6	6½	6½	7
Diameter of the hollow in inches	7	10	8	10
Length of the log in feet	15	18	21	24
Thickness after being flattened	2¼	2½	3	3½
No. of gallons boiled at a charge	50	75	120	180
No. of gallons made in 24 hours	20	30	45	60
Capacity of the steam boiler including the cylinder, which is taxable	30	45	65	75
Diameter of the condensing tube	2½	3	3½	3
No. of hours to make a run	1½	1½	2	2
Weight of the safety valve in pounds to cover or shut a 1½ inch hole	6	8	10	12
Inventor's general fee in dollars	50	90	120	150

A perpendicular view of the Log Still, with its several apartments, and the manner in which the steam passes from one to the other.

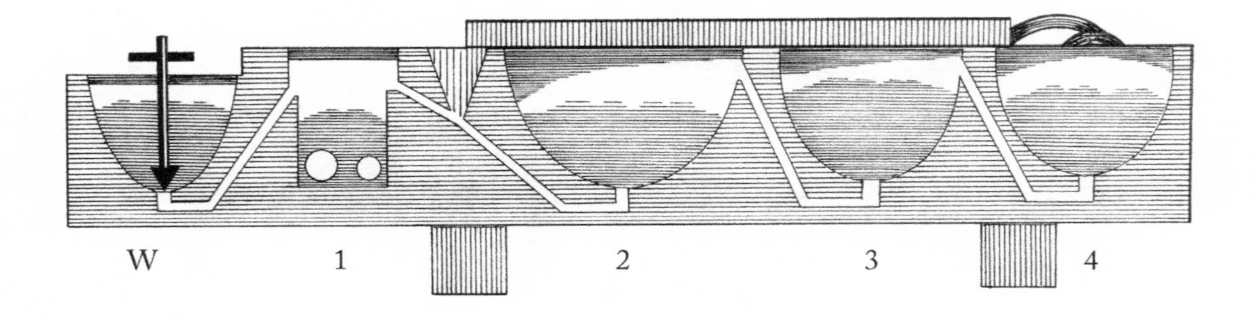

Here is shewn the shape of the several vessels, and the manner in which the steam operates on the substance when boiling, raising a curve. The caps are plain; the divisions are lined; there is a blank space under each cap, which gives room for boiling. The plank which forms one side of the beer division of the condensing cistern is represented in lines above the caps. The log rests on two blocks, the ends of which are shewn:

W designates a vessel to boil water for mashing, &c. There is a hollow cone suspended by a cord or chain from the end of a screw which works in a nut in a piece of wood across the mouth of this vessel. By turning this screw this cone is raised or lowered so as to settle in the steam passage, and being of a proper weight to resist the pressure of the steam, will when, lowered stop its passage into this vessel, and by this means the steam may be allowed to boil water for any necessary purpose, or to throw the force of the steam on the distilling vessels. By raising this cone a little the running of the still may be slackened. This cone ought to be made of lead, and its hollow may be filled with shot until of a sufficient weight. This cone also acts as a safety valve.

1, boiler for generating steam. 2 and 3, beer tubs. 4, doubler.

The quantity made per day may vary over or under that above stated, according to the strength of the materials distilling—but the above will be found as nigh as could be calculated.

The annexed drawing will give some idea of Mr. Gillespie's still. Those who wish more particular information, or are desirous of purchasing rights, will apply to Mr. Gipespie, or the author of this work.

Description of the Apparatus for Distillation, invented by Monsieur Ed. Adam.

[From the *Bibliotheque Physico Economique.*]

Chemistry for a long time possessed an apparatus, universally known, which chemists of every description used, yet no one thought of applying it in the most useful manner. The late Ed. Adam attended a course of chemistry at Montpelier during the year 7 of the French Republic. Woulf's apparatus was here exhibited to him, and he immediately conceived the idea of the possibility of applying it to the distillation of spirits of wine, on a great scale, by a single heat. He constructed a small model of his apparatus; and after numerous experiments, his patience was crowned with the most happy success. He obtained a patent for the invention for fifteen years, from the twelfth of Priarial, in the year 9, and established at Montpelier, in the department of Herault, a magnificent distillery, which attracted general notice. He thus formed a real revolution in the art of distillation; and in the year 13 he obtained a *certificate of perfection* for extracting from wine all the alcohol it contained: and on the 22d of April 1809, his brother Zachay, his heir and successor obtained another for the improvements which he added, to the invention of his brother.

This apparatus does not in any manner resemble that which was hitherto used. By the ancient process but a single species of spirit could be distilled at one heat and it was necessary to recommence the operation whenever the addition of a few degrees of strength to the spirit was desired. Adam by his new process, at a single heat, drew off at his wish, spirits of eighteen, twenty, twenty-two, and up to thirty-two degrees.

This surprising discovery excited the emulation of all the distillers, and each hastened to appropriate to himself this great invention; which was sufficiently proved by the many lawsuits for infringements of his patent, which were decided in favour of Adam.

Another process for the same purpose, founded on different principles, rivalled that of Adam. This was that of Isaac Berard.

On this occasion as well as in all important discoveries, each wished to gain for himself the honour of the invention; but it is perfectly and incontestibly known at present that Ed. Adam was the first who applied Woulf's apparatus, to the distillation of spirit on a great scale, and that the merit of the invention belongs to him. In the present state of the art of distillation (in France) sixteen different processes, which are however, but modifications and combinations of those of Adam and Berard, have given rise to the establishment of a prodigious number of distilleries over the whole empire, which produced brandy and spirits of the best flavour. The ancient processes were laid aside to give place to the new ones, and none of the old works were preserved except by the distillers of the marc of grapes, of grain, &c. either because they thought that the new processes could not be applied to their business, or feared the expense of establishing them, which is certainly considerable, but which is soon balanced, both by the *great economy of fuel* which they cause, and by the *superior quality* of the spirit which they produce, as shall be shewn, after having proved that the process of Adam is applicable to every species of distillation. Woulf contrived his apparatus to extract from various substances the gasses which they contained, and to mix with water those which are miscible with it, in order to obtain certain acids or alkalies in the liquid state. This apparatus consists of a retort, or a matrass, in which the substances are placed which are to be exposed to the action of heat.

This vessel is placed either in a sand bath or a naked fire, according to circumstances. The beak of the retort, or the orifice of the matrass, communicates with the next vessel by a bent tube, which descends to the bottom of the vessel; and from the top of this vessel, a bent tube passes to the bottom of the second vessel, and another from it to a third, and so on to any extent desired. All the vessels beyond the retort are almost filled with water, except the first, which is placed as a vessel of precaution in certain cases, which are not necessary to mention here. This is the manner in which the distillation is effected by this process. The gas disengaging itself from the retort, is carried to the bottom of the water in the next vessel; it is obliged to traverse the water in passing to the upper part of this vessel, and in doing so, a part of it combines with the water. The gas which does not combine with the water, passes by the tube to the bottom of the water contained in the second vessel; in traversing the second vessel, a part of the gas combines with the water in it, and the remainder passes into the third; and into all the following ones in the same manner. This is the principle; and the following description of the fortunate application of it which Ed. Adam made, will shew that the processes are the same.

In a furnace situated in the corner of the distillery a still is enclosed. The capital or head is in the form of a dome, and is firmly united with the body. From the middle of this dome, a tube, of the thickness of a man's arm, arises, which passes into the first vessel, placed near the furnace on a strong frame; from this vessel a second curved tube, like the first, proceeds, which passes into a second vessel, similar to the first; this communicates with a third vessel of the same kind, in the same manner. In the common distilleries, four of these vessels alone are sufficient to produce spirits of 36 degrees.

The first apparatus of Adam was more complicated than that now used: at the end of the four vessels mentioned, six others of the same kind were placed, which were called condensers, and each of which were furnished with a refrigerator. This large apparatus was only useful when alcohol extremely pure was wanted, but was dispensed with in common distilleries where spirits of 36 degrees were made.

Several essential circumstances should be observed—1st, That the vessels which are placed on the frames be all made in the form of an egg, and have their longer axes placed vertically. 2d, The entering tubes, that is to say, those which pass from the still and the vessels, into the vessels next them, should descend to the bottom of each, where they should have their extremities, in the form of the head of a watering pot, perforated with numerous holes. 3d, That the last of the egg-shaped vessels, which is most distant from the still, be furnished with a refrigerator, always full of water, during the distillation.*

Where condensers are placed beyond this last vessel, they are made exactly like it, and communicate with each other, or each separately with the first worm at pleasure, by means of cocks.

At the extremity of all these egg-shaped vessels is placed a great keeve, containing a large worm of pewter, which is surrounded with wine instead of with water, and is closed exactly in every part.

The first worm communicates with the second worm, much longer than it, which is placed in a great keeve beneath the other one, and is entirely full of water.

At the side of the great lower keeve, is placed in the earth a large cistern of masonry and cut stone, which serves for a reservoir to contain the wine that is wanted for distillation. This wine is transferred by a hand pump into the upper keeve which contains the upper worm, and all the egg-shaped vessels, as well as the still, communicate with this upper keeve, and with one another, by tubes placed in the lower parts of these vessels and of the still.

There are besides, lateral tubes, which proceed from the upper parts of all the egg-shaped vessels to the entrance of the worm in the upper keeve. Finally, small tubes pass from the upper part of all the vessels, and of the still itself, to a small worm placed in a small keeve by the side of the still.

The mode of distilling by this apparatus is equally curious. It shall be described after the manner in which all the vessels are charged the first time is related; after which the method of managing the final distillations shall be pointed out.

All the lower cocks are closed which form the communication between the great conducting tube and the egg-shaped vessels, and all the cocks of the conducting tube are opened. The wine contained in the upper keeve then passes into the copper. During this time a labourer works the pump, to restore to the keeve as much wine as had passed out from it through the tube. When the wine passes out through a little cock in the upper part of the still, it shews that the still is sufficiently charged. This little cock is shut when the still is charged, and at the same time the cock of the conducting tube, which is next the still, is closed.

The cock which communicates between the first egg and the conducting tube is then opened, and is kept open until the wine passes out through a pipe placed a little above the middle of the egg. When the small pipe is closed, and likewise the large cock, which communicates between the egg and the conducting tube, together with the cock of the conducting tube.

All the other eggs are managed in the same manner, except the condensers, where they are used, in which no liquor of any kind is placed. The refrigerators in which they are placed are alone filled with water. Then all the lower cocks are closed, and all the upper cocks are opened, to allow a free passage to the vapour, and during this time a fire is made under the still.

*The modes of distilling by steam in this country, prove that these circumstances are not essential.

When the wine is sufficiently heated to discharge alcoholic vapours, they collect in the head, and pass through the first tube to the lower part of the first egg, where they come from the tube by a number of small holes. The globules of vapour are obliged to traverse the liquor to arrive at the upper part of the egg; but it must be observed, that the vapours which come from the still are not purely alcoholic, but are mixed with much aqueous vapour. In the passage of these vapours through the wine, on their way to the empty part of the egg, the aqueous part mixes with the wine, for which it has a great affinity, and the spirituous part accumulates in the upper part of the first egg, passes from thence in the same manner to the second egg, from the second into the third, and after having traversed all the eggs, enters the upper worm where it is condensed. The refrigeration is completed in the second worm, and the liquor passes cold from the lower extremity of this second worm, and is received in a cask prepared to contain it.

When there are condensing eggs, the vapours entering into the first are partly cooled in it; the most aqueous are there condensed, and the most spirituous pass into the second, where again the most aqueous part is condensed, and in like manner successively on to the last, which transmits the most subtile vapours to the first worm, where they are condensed, before they are cooled in the second worm.

The vapours are either made to pass through all the condensers, or only through a part, according as the alcohol is required more or less pure.

It is proper to observe, that in some apparatus many of the condensing tubs are divided internally into several chambers, to cause the vapour to traverse a longer space, that none of its phlegm may be condensed.

In order that the alcohol may not evaporate in passing from the worm into the cask, and that one may see at the same time if the little streamlet of liquor runs continually, and in an equal manner, a tube is added at the end of the worm, which passes into the cask by the bung, and of which the upper part is covered with a (*or rather made of*) glass, through which the current of the liquor may be viewed.

It has been mentioned, that there was not an instance in which the alcoholic liquor, in passing from the still into the first egg, did not deposit a part of its aqueous vapour. It should not be concluded from thence, that it departed from the first egg more charged with alcohol, than when it came out of the still; that is to say, if the brandy came from the still at 18 degrees, that it ought to pass from the first egg at 19 or 20. A person reasoning thus would be deceived. A short explanation is necessary to shew that this could not happen.

The alcoholic vapour, in passing into the first egg, enters it boiling hot, and deposits in the liquor a part of its caloric, which is expended in bringing the wine in this vessel to boil. The caloric carried by the vapour into the first egg, heats the liquor, and disposes it to distillation; but this wine is not raised to the degree of heat necessary for distillation, till a long time after that in the copper has begun to distil. It is then less pure than when it entered, it has charged itself with all the aqueous vapours which could be combined with it, and which were transmitted to it at the same time that the alcoholic vapours traversed it. Two different products then rise to the upper part of the first egg, the brandy which came from the still, but freed from a part of the water it contained, and the brandy which the liquor of the first egg has produced. As this last is charged with more water than the first, it weakens the first liquor, and from this mixture a brandy is obtained which sometimes is only of the strength of 16 or even 14 degrees.

The same phenomena take place on the vapours passing into the second egg; but as the liquor is not carried to the same degree of heat as in the first egg, the aqueous vapours mix with the wine, and the alcoholic vapours rise from this second egg, mixed with a less quantity of water than those which proceeded from the first egg, and the brandy comes from it of the strength of 18 degrees. This explanation is founded on the principles of science. It is known that alcohol boils at a less degree of heat than water, consequently the alcoholic vapours are discharged before the aqueous vapours, and for the same reason, the farther any of the eggs are from the still, the more caloric have the vapours deposited in the intermediate eggs, and the less are those vapours, which come from the last eggs, charged with aqueous vapours.

When brandy is required only of the proof used in Holland, or at 18 degrees, the still and two eggs are, sufficient. Then the cock, which forms the communication between the second and third eggs is shut, and that cock is opened which admits the vapours to pass from the second egg to the upper worm, which has been before mentioned by the name of the first worm. The distillation which takes place in this case produces brandy of 18 degrees.

The products of the distillation are received in the same cask, until the liquor is perceived to be diminished in strength, then the cask is removed, and a second is applied to receive what is called the repasses (feints) in order to distil it a second time, and the distillation is continued until the still no longer affords any trace of alcohol.

When the condensing eggs are used, the liquor which collects in them is poured into the last of the distilling eggs, in order to extract from it the alcohol which it contains; and when the condensing eggs are not used, and that spirits of 35 degrees are wanted, the last egg is filled with brandy of the Holland proof instead of with wine.

To ascertain the moment when the distillation should be stopped, the first lateral cock is opened which leads to the small worm placed near the still, and that is shut which conveys the vapour from the still to the first egg; the vapours are then obliged to pass towards the small worm, condense themselves in it, and the liquor is received in a glass. Some of this liquor is thrown on the head of the still, a lighted paper is applied to it, and if it does not take fire, they conclude that the distillation should be stopped. To express this state, the distillers say that the still is spent (perdue).

The same method is used to judge of the strength of the vapours which pass from any of the eggs used: the communication between this egg and the next is stopped, leaving always a free passage between the one examined and the still; the vapours are forced to pass from it by the small lateral pipe to the little worm, and the liquor is proved either in the manner mentioned, or by the method that will be described. When the vapours which proceed from the still are no longer alcoholic, the fire is extinguished; then the discharging cock of the still is opened to let off the residue, which is of no value; it is let to run off by a pipe placed beneath the building, which conveys it to a distance.

If by the proofs previously made to ascertain the strength of the liquors contained in the eggs, it is known that they no longer contain alcohol, the cocks of communication between the eggs and the still are opened, and the liquor from them escapes along with that from the still. If on the contrary they still contain alcohol, which often* happens, then, after having shut the discharging of the still, these liquors are made to pass from the eggs into the still, in order to charge it in the same manner as at first, and the still is filled up to the proper height by adding either the feints or wine itself if necessary. The eggs are charged with wine from the first worm tub, which has been already warmed by the first distillation, which very much economises the fuel and accelerates the operations.

In the small distilleries where only three eggs are used, they make spirit of 36 degrees, by charging one or two of the eggs with brandy of 18 degrees instead of wine. When it is required to charge the eggs or the still with brandy or feints, they make use of a thick tube, which being placed between the still and the first egg, communicates with the large tube, which serves to charge the still with wine, and extends upwards as high as the middle of the eggs. A tunnel is introduced into its mouth, and in this way the liquor is conveyed into any of the vessels desired, the communication with all the other vessels being closed. When the liquor is introduced, all the cocks are closed.

Attention should be paid to one matter of consequence more. It was mentioned that the keeve filled with wine, in which the first worm is placed, should be exactly closed; but as the worm receives the alcoholic vapours very hot, the wine will be heated by them, and consequently will, as well as the eggs, disengage alcoholic vapours. The keeve is closed perfectly, in order to retain them, but least they should raise the cover, and cause the products of the distillation to escape, and to be lost, the cover is in the form of a dome, terminating in a small tube, which conveys them either directly into the worm, into one of the eggs, or into the still. By the aid of all these precautions no product of the distillation is lost.

To give a complete idea of the advantages of this mode of distilling, it should be observed, that the tube, which by the assistance of the pump conveys the wine from the cistern of cut stone to the wooden keeve, descends to the bottom of this latter keeve, and discharges itself into its lower part. The cold wine being heavier than the warm wine, occupies, always the bottom, and raises the warm wine, which serves to charge the still and the eggs. This construction presents a second advantage, which is, that the alcoholic vapours which escape from the keeve, cannot find any other vent but the tube, which conveys them into the worm or into the eggs.

In order to assist in forming just ideas of this apparatus, and of the advantages it affords, a plate of it is added, with the following explanation of its several parts.

Description of the Plate, page 61

A represents the furnace, in which the still B is enclosed to its middle part, of which only the dome or head is apparent. The dotted lines represent the form of the part of it which is concealed by the brick work. The pipe C furnished with its cock, beneath the furnace, communicates with the bottom of the still, and serves to empty it and the egg-shaped vessels. The small pipe D, furnished also with its cock, at the

*Will it not always be the case? Ed.

top of the furnace, indicates when the still is filled to two thirds of its height. A little tube E also projects from the head of the still; it is supplied with a cock, and forms a communication with the long tube X X X X, which departs from the last egg, that is to say, from the one that is most distant from the still, and passes into the small worm, immersed in the small keeve F, placed on the furnace, by which the strength of the liquors in each of the distilling vessels is proved, this small worm has a cock G at its inferior orifice.

H, H, H, a series of distilling vessels, or condensers, in the form of eggs placed firmly in the frame P Q, one after the other, at one side of the furnace.

The whole of the frame work is not represented, as it is easily conceived, it is sufficient to mention that each of the eggs is held in a frame, in which cavities are formed almost round to sustain them, so that the weight of the liquor shall not strain the frame. This frame work is supported at one side by the furnace, and at the other by the masonry which sustains the upper keeve. Only three eggs are represented here, but eight, ten, or as many as are thought proper may be added. The greater the number of the egg-shaped vessels used, the more complete will be the rectification.

The still communicates with the first egg by the tube I, which rises from the middle of the upper part of the head, and passes downward to the bottom of the egg, where it is enlarged into the form of the head of a watering-pot, pierced beneath with a number of small holes.

This tube is soldered to the egg, at the place where it enters this vessel, in order that the vapours may find no passage but that through which it is desired to direct them.

The first egg communicates with the second, and that with the third, and so on in succession, by the tube M which is soldered to the first egg at the point K, and proceeds from thence down to the bottom of the next, where it is enlarged into the form of a watering-pot head similar to the first pipe.

The last egg has a refrigerator N attached to it, by means of which the upper part of the egg, in which the vapours collect, is surrounded by water to commence the condensation. This refrigerant is furnished with a cock O, to draw off the water which it contains when it becomes too hot. When condensers are used, they are all furnished with refrigerators like this, or their lower parts are plunged into one common keeve full of water. This keeve is made of strong planks of beech, and has the form of a parallelopipedon.

The tube R serves to form a communication between the second egg and the worm, when only two eggs are used, which are sufficient to obtain brandy of 18 degrees. Then the cock of the tube M is stopped, which forms the communication between the second egg and the third, and the cock R is opened to establish a communication with the worm.

The tube S serves to form the communication of the third egg with the worm. When three eggs are used, then the cocks M and S are opened, and the cock R is closed.

The same is observed when a greater number of eggs is used; each has a tube of communication with the worm, and all these tubes are soldered to the spherical vessel T, into which the vapours pass from some of the eggs, and proceed from thence to the worm contained in the keeve U.

Distilling Apparatus by E. Adam

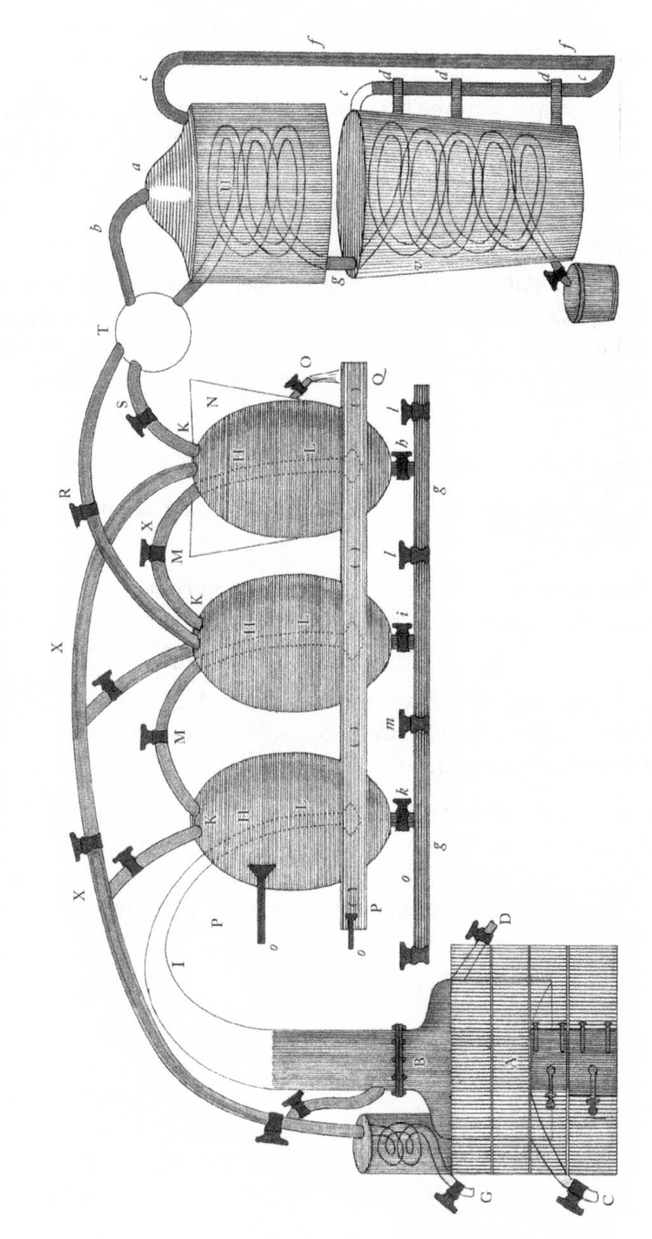

U is a keeve closely covered, which contains the first worm. It is filled with wine, which becomes warm by the passage of the vapours, which come over very hot from the first egg. It is likewise surmounted with a dome *a*, from whence departs a tube of communication *b*, which transports the alcoholic vapours, which escape from this keeve, either into the vessel T, into one of the eggs, or into the still, to pass from thence along with the other vapours into the worm. This small tube of communication is not represented here to prevent confusion in the figure, but it is easily conceived.

V is a great keeve below the first one, in which is enclosed the second worm, which is much larger than the first; it is full of water, which enters it cold through a pipe at the bottom of the keeve, while the hot water passes off above, and runs away beneath the distillery by the pipe *c* placed outside along the keeve, and supported by the three iron arms *d d d*.

It was not thought necessary to represent here the stone cistern, which serves as a reservoir to deposit the wine intended for distillation, and which a man raises into the keeve U by a forcing pump, through the conducting pipe f f f, which discharges it near the bottom of the keeve U, for the same reason stated in regard to the water which fills the keeve V.

g g g, a tube of communication between the keeve, the still, and the eggs.

h, i, k, cocks to establish, or close the communication of the eggs with the conducting tube g.

l, m, n, cocks to open, or close the communication of each egg, either with the still in order to empty them, or with the condensing keeve U to charge them.

o o o, tubes into which the brandy is poured; which is passed into them by the tunnel p, when the eggs or the still are to be charged. It is soldered to the tube g, into which it passes, and it is fastened to the apparatus by two arms, one of which is nailed to the frame work P Q, and the other is attached to the first egg.

About the middle of each egg, a small projecting tube L is placed, of the thickness of the little finger, which remains open all the time the egg is charging, and which serves to indicate when the egg is sufficiently filled. As soon as the liquor runs out by this orifice, it is closely stopped with a cork; than which probably a cock would be better.

Chapter VII.

Of the construction of Furnaces, or manner of setting up Stills.

This is an important part of the distillery business, to which sufficient attention is seldom given. Stills or boilers for different purposes, are erected in all parts of the country; fuel is generally cheap, and good workmen are rare; little or no skill is thought to be requisite in *setting* a still; if it can only be done in such a way as that it can be made to boil, the quantity of fuel is not regarded; hence few people have thought it necessary to consider the principles by which they should be guided. Independently however of the price of fuel, which is becoming in many places a matter of importance, it is necessary that the furnace be so contrived that the distiller shall at all times have perfect command of the fire, to increase or abate the heat when he thinks proper. But the increasing price of fuel renders it very worthy of attention.

Count Rumford has stated, that in general, not less than seven-eighths of the heat generated, or which with proper management might be generated, from the fuel actually consumed, is carried off into the atmosphere and totally lost!

One of the most common defects is, that of having the fire place too large; in consequence of which the bars are not entirely covered with fuel, and the cold air rushing in between the uncovered bars, counteract in a great measure the effect of part of the burning fuel. Hence, it should be a general rule to have a fire place no larger than is actually necessary to contain the fuel requisite to produce the desired effect.

The fire place should not only be small, but so constructed if possible, that the whole of the bottom of the still be exposed to action of the fire, so that the flame and heated air impinge upon the bottom before they reach the sides of the still in their passage to the chimney; for the heat which is applied to the bottom of a still or boiler, will be infinitely more effective than the same portion of heat when applied to the sides; hence the advantage of a concave, instead of a convex surface, being exposed to the action of fire.

That the chimney be furnished with an iron plate or *damper*, placed horizontally, by which the diameter of the chimney can be diminished at the distiller's pleasure, so as to moderate the action of the fire, or even stifle it at once, by shutting the openings of the fire place, and the passage of the air or smoke into the chimney.

The doors* of closed fire places are generally improperly made. They are usually made of the sheet iron, and very soon warp and become twisted, and never shut close. The mouth of the furnace should be guarded by good iron castings, with a cast iron door to fit very exactly; if the frame be *set* so as to incline a little inward, it will be better, as the door will not require a latch to keep it closed.

A good workman should always be employed, if possible; good materials are also requisite. Fire bricks should be used wherever the immediate action of the fire can be felt, and the advantage of using good bricks in the outer part of the structure is, that if it be necessary to take it down the same bricks can be used again.

The mortar which is exposed to the action of the fire, should have a very small portion of lime in it. Parkes, the celebrated chemist, advises as a cement in these cases the following mixture; good clay two parts, sharp washed sand eight parts, horse dung one part. These materials are to be intimately mixed, then beaten up with a little water, and afterwards the whole is to be thoroughly tempered like mortar, by treading it for a considerable time with the feet. This is mentioned as a suitable lute for an intense fire, and therefore may be useful in other furnaces than under stills.†

With a due attention to these rules, the following is recommended as a good

Plan for setting up a Still.

Commence the foundation sufficiently large for the circular flue round the still; build it twelve inches high, leaving in the centre an opening of fourteen inches in width, for the ash hole; then lay on cast iron pigs, so as to admit a sufficient current of air, though not of the passage of large coals; the sides of the furnace are now built up to the height of twelve inches, leaving two small flues at the back part of the furnace, two on one side and one on the other side, nearly central; the still is then set on with a good foundation of *cat and clay*, the flues being separated by a brick wall at the back of the furnace, are now carried round, the one to the right and the other to the left, until they meet in front, where they are prevented from uniting by another brick wall, this flue being half the height of the still, is now covered in, except four holes in front of four inches square, through which the flame and smoke, &c. ascend to the second flue, by which it is carried round to the chimney, the whole to be covered in when at the height of the rivets, and well plaistered.

By this means the flame or heat is carried twice round the still.

The gratings are more particularly recommended for the wash still, the boiler and doubling still may be put up without.

*For a still of 500 gallons, I found it necessary to have at the top and bottom of the door way, castings two inches thick, about 36 inches long and 14 inches wide; these were supported in front by uprights half an inch in thickness, 10 inches long and 5 inches wide, let into the top and bottom plates about half an inch; long thick straps of wrought iron were rivetted across toe door to prevent it from warping.

†A little attention to the cement used in erecting stills, will enable the operator in many instances to procure such a mixture of sand and clay as will become perfectly hard in a short time from the influence of the fire.

Chapter VIII.

Of Hogsheads, or vessels for Mashing.*

The very erroneous opinion, that casks made of any kind of wood are suitable for mashing, is generally entertained, and there are even some distillers who have adopted it. I would observe, however, as well from reason as experience, that casks made of pine, poplar, or any other soft porous wood, become so completely saturated with acid in warm weather, that it is almost impossible to sweeten them.

White oak therefore is recommended as the most suitable for making hogsheads, and the distiller commencing his operations, will find it considerably to his advantage to have casks made for the purpose; he will then have it in his power to have them perfectly smooth inside and outside, so that not a crevice shall remain to secrete any of the acid particles, at the time of scalding. They should be well hooped with iron, and made particularly strong about the chime. If however he finds it more convenient, or should be obliged to purchase casks, he must be very careful to have them well burnt or shaved inside, so that not a blister remain, for if any of these blisters should remain on the inside of the cask, a portion of the contents will insinuate itself under the blister, become acid or putrid, and cause the succeeding mash to run rapidly into the acetous or putrefactive, instead of the vinous fermentation; the produce will consequently be decreased, the quality of the spirit vitiated, and the cause will be looked for in vain.

*Hogsheads, are mentioned here, as being most generally used. Some distillers in a large way make use of casks or large tubs of 400 gallons each; these however are to be objected to on account of the inconvenience of moving them, the difficulty of scalding them perfectly, or turning them out in the fresh air, and the almost impossibility of mashing perfectly in them. It is asserted in their favour, that the greater the quantity of liquid the better the fermentation will be; this may be true in the winter season, but I rather think, that in the summer, so large a body once set to fermenting, unless cooled lower than can be done without the use of ice, which is injurious to fermentation, would proceed too rapidly for a complete vinous fermentation. I have known them tried and given up as unprofitable. Next to *hogsheads*, casks of 220 gallons, or sufficient to mash three bushels of grain, appear to me best, but I would not advise casks larger than this. I am of the opinion that vats† merely for fermenting in, might be advantageously contrived, the mashing to he done as usual in *hogsheads*, and after *cooling off* to be let down into the vats. This however could only be of consequence in very large establishments. By being placed under ground, the fermentation would not be so much affected by the weather, as is the case at present.

†It appears by the publication of Mr. Wyatt, that vats for fermenting in were used in England some years ago. These, with the mashing machine mentioned in the following chapter, fully meet my idea in the preceding note.

Of washing Hogsheads.

As soon as the fermented wash is taken out of the hogsheads, they should be washed out with cold water, the scum which has settled around the surface of the wash, carefully scraped off, and the hogshead turned out in the open air, when the number necessary for daily use, are thus washed, and boiling water ready (or the water left in the doubling still may be used for the purpose) put into each hogshead three or four buckets of boiling water, and a shovel full of hot ashes, wash them round a few times, cover them close, and let them stand about twenty minutes, then wash them out clean, and let them remain out of doors all night. The exposure to the open air evaporates any remaining acid and prevents mustiness.

This is a daily operation, and a very important one, on which in a great measure depends the goodness of the fermentations, and consequently the quantity of spirit produced. The importance of cleanliness, and the necessity of it in every thing connected with the operation of mashing, cannot be too strongly enforced; on it depends every thing; in fact no cask should be used for mashing that is not perfectly sweet, and free from the least remaining acidity.

To this point the owner or superintendent of the distillery should direct his attention every day, it is the most laborious and disagreeable operation in the business of the house, and therefore most apt to be carelessly performed by negligent hands. Milk of lime or white wash is also very effectual in neutralizing acidity and destroying must.

An additional Purifier for Hogsheads.

During the summer months the above method will not be found sufficient, of itself, to keep the hogsheads perfectly sweet; it will therefore be necessary to burn them with straw; for this purpose then, a quantity of *oat straw*, must be provided after the hogsheads are cleaned as above directed; a large handful of straw is to be put in each, set it on fire, and then turn the hogshead over on its mouth, taking care to let in just enough air to keep the fire alive, until the straw is entirely consumed. This will render the hogsheads perfectly sweet, and if repeated every time they are used, and a proper care paid to cleanliness in the other parts of the process, it will be found to improve, not only the quantity but the quality of the spirit.

If oat straw cannot be obtained, rye or wheat straw may be used, but the former is far preferable, as may be seen by making the experiment.

Chapter IX.

Of a Mashing Machine.

Various attempts have been made in England to perfect a machine for the purpose of saving manual labour in mashing grain, both for the brewery and distillery; none however appear to be successful.

A very complete one was in operation in the distillery of Mr. A. Anderson in Philadelphia, a short time since; it was worked by horse power, and the mashing at all times appeared to me to be very completely effected.

The following letter respecting a steam mashing machine, invented by Mr. Beatty, needs no comment. It is from a respectable house in New York, and the writers appear to have given it a sufficient trial to ascertain the facts stated by them.

In Mr. Anderson's distillery the mash was cooled off in the mashing tub and then let down into large fermenting tubs placed in the cellar.

New York, July 1.

"Dear Sir,

"In justice to you we conceive it our duty to state our complete conviction of the vast utility of your patent steam mashing machine, which we have now had in operation for six months, a period embracing all the possible contingencies which may naturally be expected to occur in the process of mashing; judging from this, and having had an opportunity of making trial of the usual mode of mashing here as well as seeing the most improved mashing machines in Great Britain and Ireland, we have no hesitation in declaring that it is the best mode hitherto invented for producing the greatest possible extract from grain, independently of the reduction of expense necessarily resulting from it.—With regard to the former we are quite safe in stating, from minute observations, that the extract is greater by one and a half to two quarts per bushel than that produced by manual labour: and with regard to the latter, the reduction of expense arising from men's wages alone is a consideration of great weight indeed, as will manifestly appear when we mention that, previous to the erection of the machine, we were necessitated to employ three men for the operation of mashing, now we only require one. There is a great saving of fuel attending it: the same boiler which generates steam for the stills, serving also, by means of a stop cock and steam pipe for the mash tub. As the mashing requires only the steam for about two and an half hours, it is evident the consumption of wood cannot be so great during that time as to heat a separate boiler. Upon the whole, we are decidedly of opinion that the machine will completely answer the purpose either for distillers or brewers, to whom we shall be at all times ready to give every information in our power. Wishing you therefore, all success, we remain, dear sir, your most obedient servants,

MILLER, FALCONER & Co.

Distillers, 37 Fourth-street, New York.

Mr. Leonard Beatty.

Chapter X.

The various Technical Terms used in the distillery, explained in a short account of the daily work of the house.

The first operation preparatory to mashing is to *charge* (or fill) the boiler and bring the water to a boil: then *set* the hogsheads—(put them in their proper places) a certain quantity of meal and water are now to be put in and mixed by means of the *masher* or *mashing oar*; this is made of iron bars, crossed like a gridiron, its shape round about ten inches in diameter, with a handle about five feet long; this is called *soaking the corn*; a quantity of boiling water is then to be added and stirred well, this is scalding the corn; then add the rye, and stir it well, that is mash in the rye.

The whole mixture is now called the *mash*, after it has stood some time it is *cooled off*, which is done by the addition of quantity of cold water,* then *yeast* the hogsheads, or put in the yeast. This is the operation of *mashing*. The liquid is now ready to *work* or *ferment*; and in this state as indeed in every part of the operation after scalding, it is called *stuff*.

During the fermentation it is called *stuff* or *beer*, when the fermentation is over it is said to be *ripe*, or fit to put into the still; it is now *bailed out* (i.e.) lifted up in buckets, and poured into troughs communicating with the *stand cask*, a *hogshead* in which the pump stands, to pump up the wash (as it is now called) into the *condenser* or condensing tub; the tub full, constitutes a *charge*, or sufficient quantity to put into the still; you then *charge* the still, put a fire under it, and keep constantly stirring until it boils; *the still*, is then to be *pasted* the joints to be stopped with *paste* to prevent it from *blowing*—the evaporation of the steam. As soon as the still boils, she is said to have *come round*; the liquor coming from the *worm* is called *singlings*, and according to the quantity, the still is said to run *too slow* or too fast; if the latter, there is danger of her *running foul*, that is the wash running out of the worm; to prevent this, *damp* or put out the fire, by means of a *damper*, a piece of sheet iron fixed in the chimney to stop the flue. When the singlings will no longer *burn* (to be tried by throwing a little on the still head and applying a lighted candle, or any flame) the still is then *run off*, and must be *discharged*—the contents let out into a cistern out of doors, when it is now termed *swill* or *pot ale*. The top or thin part of this after settling, is pumped up into another cistern and called *returns*, the still is then to be charged and run as before. The singlings are now put into the *doubling still*, the first running of which

*See directions for mashing.

is termed *first shot*, and must be thrown again into the singlings; you then run *spirit* until the *proof is off*, this is, when no *bubbles* remain in the bucket or receiver on the top of the liquor, the remainder is run as long as it will burn, and is called feints, and must be run over again in the next doubling.

Chapter XI.

Of Yeast.

That yeast is a necessary ingredient in commencing the process of fermentation, and of primary importance in the production of spirit from grain, affecting as well the quality as quantity, is acknowledged by all distillers both practical and theoretical, though its nature and the cause and manner of its operation, are imperfectly understood. Yet there are men, pretending to great knowledge on this subject, who will describe the various appearances which take place in fermentation, and attempt to account for them by the different kinds of yeast, and different proportions; the manner of making and of using which, is an important *secret*, known only to themselves, there is a particular drug or root, which only they know where to procure, or an *abracadabra* made use of in mixing the ingredients. By these means imposing upon the credulity of some and upon the purses of others desirous of obtaining information on the subject; retailing their secret, which many have the art of doing without detection.

Such men are not to be believed, and he who has listened with astonishment to the wonderful display of their knowledge, will be *astonished* at himself when he comes to know, that the grand secret in the art of having good yeast is *cleanliness*! Let a man make yeast according to the most approved receipts, in a dirty vessel, and he will find the produce of fermentation resulting from such yeast, far short of what it ought to be.

The importance of cleanliness in this operation cannot be too strongly enforced, nothing acid should be allowed to get into fresh yeast; and to render this easy the distiller should be provided with three small glazed earthen or tin vessels, of two or three gallons each, besides a wooden one for daily use of six or eight gallons, *and two yeast sticks*; with these he may proceed to make yeast by either of the following receipts, for the number of which an apology might be necessary, but that I wish to give a variety of methods, and because, notwithstanding *each* may occasionally prove the *best*, *each* may sometimes fail, and *all* together cannot give too much information on this important branch of the business of the distillery.

It is proper to observe that the attainment of a good stock of yeast is not so difficult as to preserve it sweet for daily use; it is therefore in the making of daily yeast that particular care is necessary; and although "feeling with a clean finger," is mentioned as a criterion to judge of heat in one of the following receipts, I would caution the distiller against putting even a *finger* into yeast, much less the whole hand as is sometimes done.

For it frequently happens that the man who is attending the stills will put his hands covered with sour wash, into the yeast. The certain consequence of which is, to sour the yeast and spoil the fermentation.

The disadvantages of sour yeast are incalculable and there is no doubt but it alone may cause a decrease in the produce of four or five quarts per bushel; even when the mashing is well done.

In summer there will be frequently anacetous tendency, scarcely perceptible however, which must be regulated by the *plentiful use of hops*.

The same stock continued long in one distillery will degenerate. Hence frequent renewals are necessary, and a change with a distant distiller will be of advantage.

1. *Of Stock Yeast.*

(From *McHarry's distiller*)

Of the following receipt the author observes, "that it will be assuredly found to contain the essence and spirit of the ways and art of making that composition, a knowledge of which I have acquired, by purchases, by consultations with the most eminent brewers, bakers and distillers in the commonwealth, and above all from a long practice and experience, proving its utility and superior merits to my most perfect satisfaction; and which I with pleasure offer to my fellow citizens, as merittng a preference notwithstanding the proud and scientific chymist, and flowery declarations or treatises of the profound theorist, may disapprove this simple mode, and offer those which they presume to be better, though they never soiled a finger in making a practical experiment, or perhaps witnessed a process of any description."

Receipt for Stock Yeast.

For a stock yeast vessel of two gallons, the size best adapted for that purpose.

Take one gallon of good barley malt, of good quality, put it into a clean, well scalded vessel (which take care shall be perfectly sweet) pour thereon four gallons scalding water (be careful your water be clean) stir the malt and water with a well scalded stick, until thoroughly mixed together, then cover the vessel close with a clean cloth, for half an hour; then uncover it and set it in some convenient place to settle, after three or four hours, or when you are sure the sediment of the malt is settled to the bottom, then pour off the top or thin part that remains on the top, into a clean well scoured iron pot (be careful not to disturb the thick sediment in the bottom, and that none goes into the pot) then add four ounces good hops, and cover the pot close with a clean scalded iron cover, and set it on a hot fire of coals to boil, boil it down one third, or rather more, then strain all that is in the pot through a thin hair sieve (that is perfectly clean) into a clean well scalded glazed earthen crock, then stir into it, with a clean stirring stick, as much superfine flour as will make it about half thick, that is neither thick or thin, but between the two, stirring it effectually until there be no lumps left in it. If lumps are left you will readily perceive that the heart or inside of those lumps will not be scalded, and of course, when the yeast begins to work, those lumps will sour very soon, and of course sour the yeast; stir it then till all those lumps are broken, and mixed up, then cover it close for half an hour, to let the flour, stirred therein, be properly scalded, after which uncover and stir it frequently until it is a little colder than milk warm (to be ascertained by holding your finger therein for ten minutes, but *beware your finger is clean*) then add half a pint genuine good yeast (be certain it is good, for you had better use none, than bad yeast) and stir it effectually until you are sure the yeast is perfectly incorporated with the ingredients in the pot; after which cover it, and set it in a moderately cool place in summer, until you perceive it begin to ferment, or work, then be careful to stir it two or three times at intervals of half an hour—then set it past to work—in the winter, place it in a moderately warm part of the still house, and in summer, choose a spring house, almost up to the brim of the crock in water—avoiding extremes of heat or cold, which are equally prejudicial to the spirit of fermentation—of consequence, it should be placed in a moderately warm situation in the winter and moderately cool in the summer.

This yeast ought to be renewed every four or five days in the summer, and eight or ten in the winter but it is safer to renew it oftener, or at shorter intervals, than suffering it to stand longer. In twenty-four hours after it begins to work, it is fit for use.

Between a pint and half a pint of the foregoing stock yeast is sufficient to raise the yeast for the daily use of three hogsheads.

2. *Mode of separating Beer from Yeast, and preserving the Yeast for a great length of time in any climate.*

Mr. Felton Matthew, merchant, London, obtained, a patent for the above mentioned object, which may be found in the *Repertory of Arts*, vol. v. p. 73. Mr. Matthew uses a press with a lever, the bottom made of stout deal, oak, or any other timber fit for the purpose, raised with strong feet a convenient height from the ground, so as to admit the beer to run off into whatever is prepared to receive it. Into the back of it is let a strong piece of timber, or any other fit material to secure one end of the lever, the top of which is secured by being well wedged up to a girder, or the joints at the top of the building. In this piece of timber is mortised one end of the lever, which is fastened into the mortise with an iron pin, or otherwise properly secured; the whole well secured with iron work. The yeast is then put into bags made of sail cloth, or any other strong cloth or materials, and carefully tied or secured, then placed flat on the press; a board is then laid on it and the lever let down on it, and weights are hung at the other end of the lever, by hooks or otherwise, and weights are added as the beer runs from the bags, care being taken not to burst the bag nor to force the beer out too thick; which to prevent, the bag is placed in a trough of a proper size, with a false bottom, bored full of holes (the sides and ends being likewise bored full of holes) and blocks put above for the lever to act upon. When a sufficient weight has been added to force the beer out completely, which may be done by a screw press, if necessary, the yeast, which remains in the bag will crumble to pieces. It must then be thinly spread upon frames made with canvass, hair cloth, or any other thing which will permit the heat to pass freely through it, in a room, kiln or stove, or other place where a regular heat can be kept up to the temperature of from about eighty to ninety degrees; observing to break it fine as it dries, by passing a board or other fit thing lightly over it. When completely dry, put it into tight casks or bottles, so as to exclude the air or any damp, from it, and it will then keep a great length of time, and in any climate. When wanted for use, it may be dissolved in a small quantity of warm wort, or sugar and water of the temperature of eighty or ninety degrees, when it possesses the same quality as fresh liquid yeast.

The above directions are copied at full length for the benefit of those who may wish to put up large quantities of yeast; some distillers being of opinion that this is the best mode of keeping a constant supply of sweet stock yeast. The principle being established, few distillers will be at a loss for a method of carrying it into operation. In the country there are few farmers without cheese presses which will answer the purpose perfectly well.

3. *Substance of the specification of a patent granted to Mr. Richard T. Blunt; for his new invented composition to be used instead of yeast.*

To make a yeast gallon of the above mentioned composition, containing eight beer quarts; boil in common water eight pounds of potatoes, as for eating; bruise them perfectly smooth, and mix with them whilst warm two ounces of honey, or any other sweet substance, and one quart (being one eighth part of a gallon of yeast) of common yeast.

Remarks by H. H.

The above may be used without using any yeast as follows. Prepare the potatoes as above, put a small handful of hops into three pints of water and boil it down to a quart, strain it and let it cool to 150 degrees, then add a pint of malt, stir it well, and mix with the potatoes when about milk warm; add a half pint of honey or molasses, cover and let it stand in a warm place; it will be fit for use in six or eight hours.

4. *A substitute for brewers Yeast, patent!!*

Take six pounds of malt, and three gallons of boiling water, mash them together, cover the mixture and let it stand three hours; then draw the liquor off, and put two pounds of brown sugar to each gallon of liquor, stir it well till the sugar is dissolved; then put it in a cask just large enough to contain it, and cover the bung-hole with brown paper; let it stand four days, kept to a blood warm heat. Prepare the same quantity of malt and boiling water as before, but without sugar, mix it all together and let it stand forty-eight hours, when it will be fit for use. This is called by the patentee, *the fermentation.*

To make seventy-six gallons of the substitute.

Put twenty-six ounces of hops to as many gallons of water; boil it full two hours so as to reduce the liquor to sixteen gallons. Take this and mash it with the malt when the liquor is at 190 degrees; it must now stand two hours and a half, and be strained; ten gallons of boiled water, at the same heat, is to be mashed with the malt, strained and cooled. Take the first liquor when blood warm, and put to it four quarts, of the fermentation: mix it well, and let it stand ten hours. Take the remaining ten gallons of the liquor, and put it with the sixteen gallons of liquor, let it stand six hours, and then it is fit for use, in the same manner and for the same purposes which brewers yeast is made use of.

The advantages attending this invention are, that the substitute for yeast will keep sweet and good longer than brewers yeast, may be used in all weathers and climates, and is the means of making bread more white and lighter than brewers yeast.

5. *For stock Yeast, another receipt.*

Take as many hops as may be held between the thumb and three fingers, put them into a quart of water and boil them well together, a few slices of apple or pumkin put into it will be an advantage. Then pour the liquor off, or strain it through a coarse cloth, and add three spoonsful of molasses, and stir in as much wheat flour as will make it of the consistence of batter, cover it over and set it in a proper temperature and it will be fit for use in seven hours.

6. *Another patent Yeast for keeping.*

Half a quarter of hops put to one gallon of water, and boiled to a quart, when boiling hot strain it, and thicken it with rye flour, quite stiff, when sufficiently cool add a quart of good yeast, let it rise twenty-four hours; then take fine boulted Indian meal and thicken it until thick enough to mould out upon a table; roll it, cut it in slices, dry it perfectly, then put it up for use. To be made in cool weather.

N.B. If good stock yeast be well strained and mixed with molasses in the proportion of one to three, it may be kept for a long time in well corked bottles.

7. *To preserve a good stock of Yeast, for the summer season.*

About the latter end of the month of April, take two or three gallons (or more if very extensively engaged in business) of good sweet yeast, that is, in a perfect state of fermentation; thicken it with coarse wheat middlings or chopt rye, adding at the same time about two gills of whiskey. Spread it out upon a clean board, and after it has dried in the sun a few hours, rub it through your hands so as to break up all the lumps: expose it daily to the sun and air until it becomes perfectly dry, taking care not to leave it out until the dew begins to fall; it is then to be put up in small paper bags, and kept in a dry place until wanted for use.

By a careful attention to these directions, the distiller may rely upon having a perfectly pure and sweet stock of yeast which will remain so for any length of time, if kept in a dry place. About two hours before you wish to make use of dried yeast, take three or four gills, put in a proper vessel, and pour thereon a sufficient quantity of warm water to make it of the thickness of mush, stir it well, and in a short time it will be in a state of fermentation, fit to raise the necessary quantity of yeast for daily use.

A sufficient quantity of the above yeast should be made in the spring to last until fall, calculating upon the use of about one pint per week, as it will generally be found necessary to renew yeast once a week in warm weather. It would also be well enough to put away a small quantity in the fall for use during the winter, when a renewal every three or four weeks will be found of great advantage, and sometimes actually necessary.

8. *Of Yeast for daily use.*

A sufficient number of receipts being given to put the distiller in possession of a good stock of yeast, and to enable him to preserve it sweet and good, it now remains to apply it to daily use. For this purpose, care should be taken, that not only the water and the vessels to be used, are perfectly clean and sweet, but that the meal to be used should be entirely free from lumps, mustiness, or acidity; for, upon a proper attention to this part of the process depends in a great measure the success of the distiller.

9. *For the Yeast necessary for eight hogsheads.*

Have a wooden tub capable of holding six gallons, take a handful of hops, rub them between your hands so as to separate the leaves, into the tub, pour on them four gallons of boiling water, cover it with a cloth, let it stand a few minutes until the heat is about 155° of Fahrenheit, then stir in malt and rye meal mixed, in the proportion of one fourth of the former and three fourths of the latter, stir it until there are no lumps left, let it stand covered for about fifteen minutes, then uncover and stir frequently for half an hour. It is then to be mixed by small quantities with the stock yeast, in another vessel, taking care not to mix it too warm, nor so much at a time as to retard or stop the working of the stock yeast. It must be so managed, as that the whole shall be mixed in this way about an hour before cooling off, except about one gallon, which is to be reserved until you cool off. Then add to it about one pint (*more or less*, according to the weather) of yeast, and set it away for stock the next day. It will be found in a fine state of fermentation. This is the best way to keep yeast from day to day, and with attention will be sweet until the stock degenerates.

According to these two last receipts, yeast was generally managed in the author's distillery.

To judge of the quality of Yeast, and to sweeten it when necessary.

Observe if it works quick, sharp and strong, and has increased considerably in bulk, at least one half more than before it commenced working; that it has a sweet taste and smell, with somewhat the appearance of a honey comb, constantly changing its place, and rolling like waves from the sides to the centre, colour bright and lively, and the head constantly rising, or rather, not having fallen from its greatest height; these appearances indicate good yeast. But on the contrary if it be dead or flat, works sluggishly, *head* considerably fallen, and has either sour taste or smell, it must be renewed, or if there be not sufficient time for this, before the yeast is wanted, it may be sweetened by simply scraping into it a *small quantity of chalk.*

This however is the last resource of laziness, and die good distiller should be careful not to be obliged to recur to it.

Account of a method of generating Yeast.

[Communicated by the Rev. Wm. Mason, to the Society for the Encouragement of Arts, Manufacture and Commerce. A bounty of £20 sterling was given to his servant for the discovery.]

This experiment was first made by impregnating wort with fixed air, according to, the ingenious method recommended by D. Henry, and succeeded very well. Another experiment was made without fixed air, which also succeeded;—a small quantity of yeast was made with no other ingredients than malt, water and heat. The original quantity made was increased in a few days, until it became sufficient to work a hogshead of small beer, which produced ten pounds of perfect yeast, and this being soon after put into a vat for a hogshead of ale, was augmented to forty-two pounds.

The discovery therefore is simply this: "That yeast is not (as has been thought) some peculiar and unknown substance, necessary to be added to wort in order to put it into a fermenting state; but that malt boiled in water will generate it (as the chemists say) *per se*, if the following circumstances be attended to:

1st. That the process be begun with a small quantity of the decoction.

2d. That it be kept in an equal degree of heat. And,

3d. That, when the fermentation is begun, it should be assisted and augmented with fresh decoctions of the same liquor."

For the proportions and method found to succeed by his servant, Dr. Mason gives this

Recipe:

Procure three earthen or wooden vessels of different sizes and apertures, one capable of holding two quarts, the other three or four, and the third five or six: boil a quarter of a peck of malt for about eight or ten minutes in three pints of water; and when a quart is poured off from the grains, let it stand in a cool place until not quite cold, but retaining that degree of heat which brewers generally find to be proper when they begin to work their liquor. Then remove the vessel into some warm situation near the fire, where the thermometer stands between 70 and 80° (Fahrenheit) and there let it remain till the fermentation begins, which will be plainly perceived within thirty hours; add then two quarts more of a like decoction of malt, when cool, as the first was, and mix the whole in the larger sized vessel, and stir it well in, which must be repeated in the usual way, as it rises in a common vat, then add a still greater quantity of the same decoction, to be worked in the largest vessel, which will produce yeast enough for a brewing of forty gallons.

This has never failed, except when the failure was to be accounted for from an inequality of temperature of the air, where the experiments were made; therefore, in the hands of a good practical brewer, accommodated with a place where his little vat will stand in a constant degree of proper heat, it will generally succeed, especially in brewing seasons.

Dr. Mason adds, that the experimenter was of opinion, that a proper quantity of hops boiled in the liquor makes the fermentation proceed better, but as it may, and has actually succeeded without such addition, it had better be omitted where the yeast is wanted for bread, lest it make it bitter.

Dr. Mason defines yeast, used for the purpose of brewing malt liquor, or raising bread, to be a "viscid frothy substance which arises on the surface of a simple decoction of malt in water, when in a state of fermentation, and which substance, after it has been so generated, may, by additional quantities of the same liquor, gradually supplied, be increased *ad infinitum.*"

For the purpose of keeping up an equable temperature around the vessels containing the yeast, Dr. Mason made use of a small box, about 12 or 14 inches square, open on one side, the open side being placed fronting a warm wall or stove, or fire-place. This heat, he says, should be rather below 80° of Fahrenheit. In this box he made the following:

Experiment 1. Three vessels were set at the same time in the warm box, containing a quart of liquor each, and of equal strength with respect to malt; one was a decoction without hops, another with hops, the other a pure infusion of malt; in about twenty-four hours the hopped decoction produced a fine head of yeast; the other decoction fermented as well, but was twenty-four hours later; the simple infusion was near thirty-six hours later, and the yeast appeared dark, and ill coloured, so that it was thought to be spoiled: its bad appearance however was only owing to its not having been boiled and cleared, for it made very light breakfast rolls.

This experiment, the Doctor concludes, proves that hops *accelerate** the fermentation; it would seem also to prove, that neither hops of boiling were *necessary* to the process.

Experiment 2d. Four vessels from a common brewing of all were placed in a box of longer dimensions; one contained two quarts; a second, one quart; a third, a pint; a fourth, half a pint: they all shewed signs of fermentation the same time, viz. in about twenty-four hours; but that in the mug or pot holding a pint appeared the strongest, which was thought to be owing to the smaller diameter of the vessel, which was smaller in proportion to the half pint; but as it stood more centrally to the heat of the fire behind, I am persuaded the excess of fermentation proceeded from that cause. This proves, that the quantity with which the process is commenced, is not material.

Experiment 3d. Was instituted merely to find whether an addition of sugar would accelerate the fermentation; for which purpose, two quarts of hopped liquor were tried in separate vessels, a quart each, and the result was, that the decoction in which two large spoonsful of cane sugar were stirred in, *did not ferment in the least*, though continued in the warm box five days and nights; the other fermented in about thirty-six hours. The reason of this later fermentation than that of the others in the former experiment was, that the liquor used was from a brewing of small beer. Hence we may conclude, that a decoction of the strength of ale, if not of strong beer, is the best to begin with.

*This is contrary to the general opinion, that hops retard and regulate the fermentation.

Chapter XII.

Of Mashing.

A sufficient number of hogsheads being had in readiness, and a good stock of yeast prepared, agreeably to the directions in the preceding pages, the important operation of mashing next requires attention.

This process, upon the correct performance of which, the success of the distiller greatly depends, is by too many considered as a mere mechanical operation, and hence arises the very great difference of opinion among distillers, and the variety of results which are observed even in the same distillery. Why should there be this difference in a matter which it is in the power of every operator to reduce to a tolerable degree of certainty? Prejudice does much, but ignorance does more. Our distilleries were for a long time conducted by men ignorant of the principles of the art, and entirely unacquainted with the causes which produced the certain effects; a particular method was adopted which was transmitted from one to another without the least deviation. Heat, that all-powerful agent, upon the proper application of which success entirely depends, was scarcely attended to, or judged of by feeling with the fingers, than which, a more fallacious method could not be devised. Hence the various results, and consequent surprise of the ignorant operator.

To guard against this variety in the results, and render them as uniform as possible, particular attention should be paid to the precise quantity of ingredients employed, and the heat regulated by a thermometer, an instrument absolutely necessary in every distillery. And the young distiller should be careful to make notes of every part of the operations, and the results, by a comparison of which, he may be enabled to regulate his future operations with a greater degree of certainty.

The term *mashing* in its usual acceptation, may be applied to any mixture of meal and water, at any degree of heat, or *the art of mixing*. Among distillers however, it would be more properly defined to be that process by which meal or grain is prepared for vinous fermentation, by a skilful mixture of certain portions of meal and water at a particular degree of heat, which being continued, separates and dissolves the component particles of the grain, developes the saccharine matter, and prepares it for entering at once into the vinous fermentation.* Hence the very essence of the process

*This ought to be considered as the first stage of the vinous fermentation, and the malt as the ferment which induces it. Hence it is, that when the malt is put in at the same time with the Indian corn, the sweetening and cracked appearance take place almost as soon as the hot water is put on them, and grain mashed in this manner, may be cooled off to 72 or 75° in summer, and proportionally low in winter; because of the strong

is, the proper application of heat, according to which will the process be more or less complete, as is abundantly shewn to every one who has paid due attention to it, by the quantity of saccharine or sweet matter which rises to the top of the *mash*, when the proper degree of heat has been applied.

This precise degree of heat can only be attained, with certainty at all times, by the use of the thermometer; but as this instrument cannot always be had, the operator must then judge of the mash from the appearance and taste.

When Indian corn (maize) is properly scalded, upon stirring it well, and letting it stand for a few minutes, it will be found cracked in several places, and have the appearance of water oozing from the cracks, the mashing oar moves freely; there will be no lumps, and upon withdrawing the mashing oar, but a small portion of the mash will adhere to it, and will appear transparent. If too highly scalded none of these appearances are perceptible; the mash is thick and lumpy, appears dead and heavy, and adheres very closely to the mashing oar, which it is difficult to move, and the taste is like well boiled mush.

When not sufficiently scalded, the appearance is very watery, or like cold water and meal mixed, and upon standing a few minutes the meal separates and sinks to the bottom of the cask, leaving the water clear, and with the taste of raw meal.

When rye is well scalded, the same appearances occur as in Indian corn well scalded; it is transparent, and but little adheres to the mashing oar; the colour is somewhat increased in browness, and upon standing, a white cream rises to the top of the mash, which is *perfectly sweet*; this takes place within 20 or 30 minutes after the mash is made, and according to its quantity may the perfection of the process be judged of, and the produce pretty accurately estimated; this is a sure indication of a good mash, and appears with all kinds of grain.

If not sufficiently scalded the raw taste is apparent, and boiling water may be added to raise the temperature; but if too highly scalded, the mash is very lumpy, thick and clammy, adheres closely to the mashing oar, no sweetness is perceptible, the fermentation will be bad for want of a sufficiency of saccharine matter, and upon distillation, will be liable to adhere to the bottom and sides of the still, and burn. There is no remedy when grain is *over scalded*, therefore this should be most carefully guarded against; but when any kind of grain is not sufficiently scalded, hot water can be added, and sometimes produce the desired effect.

Another thing to be attended to is, always to use boiling water, the temperature of which being known, can be reduced to any desired degree by the addition of cold water; whereas, by taking water from the boiler, *supposed* to be of a proper heat, you will be involved in uncertainty, and no two casks will be mashed at the same degree of heat, consequently will be wrong.

The mashing room should be kept as nearly as possible at one temperature, which may be easily done in winter; the best heat will be from 65 to 70°. The mashing

disposition to vinous fermentation already imparted by the malt. Having thus ascertained the use and effect of the malt, we see the advantage of putting it in at the very beginning of the process, and we also see the reason why we have always been disappointed in expecting any benefit from adding it just before the time of cooling off.

floor should be perfectly solid, that when hogsheads begin to ferment, they should not be disturbed by persons walking through the house. For, although weak and languid fermentation may be assisted by briskly agitating the fluid, yet absolute rest is necessary to obtain the slow and regular vinous fermentation, which is most productive of spirit.

It is the practice of some distillers to *cool off* two *hogsheads*, with one already fermented and fit for the still; in which case no yeast is used.

By thus concentrating the grain of three hogsheads into two, there is a saving of room in the house, and of labour and fuel in distillation. But it is directly contrary to the experience of many good distillers, and to the theory of chemists, who say "that matter once fermented yields, by a second fermentation, acid alone."

Hence, where a complete vinous fermentation has taken place, a second fermentation of the same matter must produce vinegar.

Experiments in distilleries are generally made in so careless a manner as not to be depended upon, but this mode being persevered in by some distillers renders it worthy of investigation.

The results of my experiments always were, that two casks fermented with a third, produced rather more spirit than two casks fermented in my usual way, but much less than three.

Hence there was always a loss, but not of the whole of the third cask.

I would strongly recommend the keeping all acidity from casks, previous to and during fermentation. But in very cold weather it may be sometimes adviseable, at cooling off, to add a few gallons of wash, in a high state of fermentation, to each cask.

The returns will be found to have lost much of their acidity by passing through the still, therefore are not liable to the above objections.*

To mash forty-five pounds Corn, and forty-five Rye and Malt.

Put into a hogshead four gallons boiling and four gallons cold water, or in such proportion, according to the weather, that the heat may be about 110°; let it stand a few minutes, then stir in 45 pounds Indian meal, cover the hogshead, and let it stand from one and a half to two hours, according to the weather; then add 12 to 16 gallons boiling

*In opposition to this reasoning and theory may be mentioned, the mode of mashing adopted in Tennessee and Kentucky, to wit: of *mashing with pot ale*. It is not uncommon in some of the distilleries in those states to use dirty casks, into which the requisite quantity of pot ale in a boiling state is thrown, the corn meal is then added, and well stirred; in this state it is suffered to stand three, four or five days, when a small portion of rye meal and malt is added, and the whole is cooled off. *No yeast is added*, and the stuff is ready for the still in about four days more. It is somewhat singular, that any spirit should be extracted from the rye mashed in this way, the temperature probably not exceeding 125°; yet, in many experiments casks mashed in this manner, yielded as much as others where the rye was mashed separately in the usual way and added. We need not argue against facts, and the certainty that these things are so, only serve to shew us the folly of any one insisting upon his mode of mashing being decidedly superior to any other, when similar results are obtained by so many different processes. It is by collecting, comparing and analysing these facts, that we may hope to arrive at the great desideratum in mashing.

water, stir it well, and add 12 gallons more of boiling water, so that the heat upon its being well mixed, may be from 155 to 160°; let it stand covered for 20 or 30 minutes, then add 35 pounds rye meal, and 10 pounds of malt, stir it well until all the lumps be broken, cover it for 20 minutes, after which uncover and stir it frequently until it is fit to cool off, which will be in winter about three hours, and summer five or six.

It should be cooled off in winter with 25 gallons returns, and water sufficient to make it about 80 or 90°, in summer 75° if possible, and fill the cask within six inches of the top; add one or two quarts of yeast, cover it, and it will be fit for the still in 50 or 60 hours.

The following receipt is said to be the most approved Russian method.

Put 10 gallons boiling water into the mash tub, add 40 pounds corn, mash it well, pour on two gallons more boiling water and stir it, then put on 20 gallons more of boiling water; put in 45 pounds rye meal, stir it well, and strew over the mash five pounds meal (or malt), let it stand closely covered for six or seven hours, then cool it off with water down to 90°; add one quart of yeast, cover the tub, and in about 60 hours it is fit for the still.

To mash two-thirds Corn and one-third Rye.

The process is the same as in mashing equal quantities of each article, except that a little more water will be necessary to *soak* the corn in the first instance, and always bearing it in mind that the greatest heat necessary in scalding corn, should rather be below than above 160° of Fahrenheit.

To mash Indian Corn alone.

This should be avoided if possible, the weight of corn being very great, and having a very small proportion of bran, it sinks to the bottom of the hogshead and is consequently difficult to be fermented. Therefore it will be always to the advantage of the distiller to mix some other grain with it.

Process.

Put into a hogshead 12 gallons water of about 110°, then add 70 or 80 pounds corn, stir it well and let it stand from one to two hours, then add 24 gallons boiling water, stir it well, cover it for 30 minutes, then add the malt and stir it frequently, until fit to cool off as above. It must be cooled off warmer than where there is a mixture of other grain, requires more yeast, and about 25 gallons of returns in each hogshead, except in very warm weather.

N. B. I have always considered the use of the returns as very important, and have observed that they cause an increase in the yield of about one-tenth. It should however be observed, that by persevering in constant use of returns, they become of little effect. They should therefore be occasionally omitted for a week, so that a fresh stock may be brought into use,

The following is the method of mashing Rye, pursued in some of the first Distilleries in Holland.

Put into a hogshead 30 gallons of water heated to 165°, add to it 20 pounds of malt finely ground, stir it well, and let it stand 15 minutest, then add 70 pounds rye meal, and stir it well, when completely mixed, the heat of the mash should be from 145 to 150° Fahrenheit. Let it stand three or four hours according to the weather, stirring it frequently, then add 20 to 25 gallons of returns that has stood over night, and is clear, fine and cool, stir it in and fill up with clean water, so as to reduce it to a proper heat, then yeast it, and it will be fit for the still in 48 or 60 hours.

I have somewhere met with the following as the Holland plan of mashing.

"Take ten pounds malt, ground very fine, and three pounds of common wheat or rye meal, add two gallons cold water, and stir all well together; then add five gallons boiling water. When this is *cold*, add two ounces *solid* yeast, and ferment it in a warm place loosely covered."

It is not unusual to vary these receipts as to the time of putting in the malt; sometimes to mix the malt with cold water, or with the water at 110 or 120°, and let it stand from ten to twenty minutes before the corn is stirred in; neither of these methods ever appeared to me to make any difference in the product. It is proper however to mention, that the *cracked appearance* takes place sooner when the malt is mashed first; but then the mash cannot be so well judged of by the taste, as the malt always sweetens it. The quantity of malt may be increased or diminished according to circumstances.—*See Malt.*

To mash Rye.

(See *M'Harry's Distiller.*)

"Take four gallons boiling water, and two gallons cold water, put it into a hogshead, then stir in one and a half bushels chopped rye, let it stand five minutes, then add two gallons cold water, and one gallon malt, stir it effectually; let it stand till your still boils, then add sixteen gallons boiling water, stirring it well, or until you break all the lumps, then put into each hogshead, so prepared, one pint coarse salt, and one shovel full of hot coals out of your furnace (the coals and salt have a tendency to absorb all sourness and bad smell that may be in the hogshead or grain; if there be a small quantity of hot ashes in the coals it will be an improvement; stir your hogsheads effectually every fifteen minutes, keeping them close covered until you perceive the grain scalded enough, when you may uncover; if the above sixteen gallons boiling water did not scald it sufficiently, water must be added until scalded enough, as some water will scald quicker than others; it is necessary to mark this attentively, and in mashing two or three times, it may be correctly ascertained what quantity of the kind of water used will scald effectually , after taking off the covers they must be stirred effectually every 15 minutes till you *cool off.*

To mash two-thirds Rye and one-third Corn in summer.

"This I have found to be the nicest process belonging to distilling; the small proportion of corn, and the large quantity of scalding water, together with the easy scalding of rye, and difficulty of scalding corn, makes it no easy matter to exactly hit the scald of both.

"Take four gallons cold water, put into a hogshead, then stir half a bushel of corn into it, let it stand uncovered 30 minutes, then add 16 gallons boiling water, stir it well, cover it close for 15 minutes, then put in your rye and malt, and stir it until there be no lumps, then cover and stir it at intervals until your still boils, then add 8, 12, or 16 gallons boiling water, or such quantity as you find from experience to answer best (but with most water 12 gallons will be found to answer) stirring it well every 15 minutes until you perceive it is scalded enough, then uncover and stir it effectually until it is fit to cool off; keeping it in mind always, that the more effectually you stir it, the more whiskey will be yielded. This method I have found to answer best; however, I have known it to do very well, by soaking the corn in the first place with two gallons warm, and two gallons cold water, instead of four gallons cold water mentioned above; others put in rye when all the boiling water is in the hogshead, but I never found it to answer a good purpose, nor indeed did I ever find much profit in distilling rye and corn in this proportion.

"To know when rye is sufficiently scalded, take up some of it on your mashing stick, and you will perceive the heart or seed, of the rye, like a grain of timothy seed, sticking to the stick, and no appearance of mush."—See *M^c^Harry's Distiller.*

The two last receipts are published in conformity to the plan of giving a variety, out of which the reader may have a choice, and because the writer of them says, they are the best that could be obtained. I would recommend as absolutely necessary however, the use of perfectly sweet casks instead of the "salt and coals," mentioned to take away the acidity after mashing; they will not have the effect.

Previous to the publication of the former edition, a respectable friend advised the insertion of a few experiments in mashing, or a particular account of that process every day, with the various results. Although I have at all times kept memorandums of my work, yet those which I then had, were not sufficiently accurate for the purpose. I have therefore endeavoured to remedy the deficiency by the following account of the daily work for 38 days during the winter of 1814–15.

It will be observed, that for a number of days the weather was very cold, the thermometer being but a few degrees above the freezing point in the middle of the distillery, the produce was consequently not such as it ought to have been, and not equal to the produce of similar mashing when the thermometer stood higher.

For several days I used a process recommended to me of mashing entirely with boiling water. The produce was never equal to eight quarts per bushel. I mention this to warn others from a similar error; for it is frequently as beneficial to make known the failure of an experiment, as to publish a successful attempt.

Monday, Dec. 5, 1814—Thermometer 40°— Mashed 7 hhds. 4 gal cold water 56°. 6 gal boiling water, 60 lbs corn meal, well mixed and stood for one hour; then 16 gal boiling water, heat 148°, then 8 gal do. heat 165°, stood 20 min; then 2 lbs malt and 25 lbs rye meal. Cooled off in two hours and a half, at 3 o'clock, PM 84 *a* 86° 3 qt yeast, fermentation slow. At 8 AM on Tuesday, head not broke.

Same day, 6 hhds. additional. 60 lbs corn meal soaked as above, stood 1 hour 30 min, then 16 gal boiling water 146°, then 8 gal do. 160°, stood 15 min, then 2 lbs malt and 25 lbs rye. Cooled off in two hours and a half, at 7 PM 86 *a* 90° 3 qt yeast.

Tuesday, Dec. 6—Therm. 34°—6 hhds. Corn soaked as usual, stood 1 hour 45 min, then 16 gal and 8 gal boiling water 158°, stood half an hour, then 2 lbs malt and 25 rye. Cooled off in two and a half hours, at 4 PM 90 *a* 92° 4 qt yeast.

Fermentation on Wednesday, at 7 AM *only tolerable*. Yeast for these 19 hhds. 28 lbs rye.

Wednesday, Dec. 7—Therm. at 7 AM 36°—6 hhds. Corn soaked as usual, stood 1 hour, then 16 gal boiling water 150°, stood 10 min, 8 gal boiling 158 *a* 161°, stood 20 min, then rye and malt. Cooled off in 3 hours, 1 o'clock, 86 *a* 88°.

Same day. 6 hhds. Corn soaked as usual at 10 hours 30 min, stood till 12 hours 5 min, then 16 gal boiling water 150°, at 12 hours 20 min, 8 gal do. 163°, stood till 1 o'clock, then malt and rye. Cooled off at 3 hours 30 min, 86 *a* 89° 3 qt very good yeast. These casks run 30 gal singlings per charge.

Thursday, Dec. 1—Therm. 7 AM 34°—6 hhds. Corn soaked as usual at 8 hours 45 min, in one hour 16 gal boiling water 150°, 8 gal do. 160°, at 13 hours 15 min malt and rye. Cooled off at 1 o'clock, 86 *a* 89° 3 qt very fine yeast. Head raised at 7 PM.

Same day, 6 hhds. Corn soaked as usual at 2 hours 30 min, at 3 hours 45 min added 16 gal boiling water 150°, then 8 do. 159°, at 4 hours 15 min malt and rye. Cooled off at 6 hours 45 min 86 *a* 90° 3 qt fine yeast.

At 7 AM on Friday, no head or visible fermentation on 2 hhds. which however produced their equal proportion of singlings.

Friday, Dec. 9—Therm. 7 AM 36°—7 hhds. Corn soaked as usual at 9 hours 20 min, at 10 hours 20 *a* 30 min 16 gal boiling water 151°, then 8 do. 163°, stood 20 min, then malt and rye. Cooled off at 1 hour 30 min, 86° 3 qt fine yeast. No head on 4 hhds. next day at 5 PM owing to cold weather; added 4 gal boiling water and 2 qt yeast to each.

Saturday, Dec. 10—Therm. 7 AM 32°—7 hhds. As above. Cooled off at 3 PM 90° therm. 42°. Run 4 charges of wash 2 hhds. each, produce 24 gallons singlings per charge.

Monday, Dec. 12—Therm. 7 AM 32°—7 hhds. As above. Coooled off at 3 PM 32°. 3 basins very good yeast. No head the next day at 7 AM.

Tuesday, Dec. 13—Therm. 7 AM 40°—7 hhds. Corn soaked, as usual at 8 hours 20 min, stood till 9 hours 40 min, then 16 gal boiling water 145°, 8 do. 153°, stood 10 min, then rye and malt. Cooled off at 1 hour 20 min, 82 *a* 86° 3 basins very fine yeast. This day began to use rye and malt at the rate of a half bushel to 7 hhds., for yeast.

Same day, corn soaked as usual at 11 hours 20 min, stood 1 hour, then 16 gal boiling water 148°, 8 do. 159°, rye and malt as above at 1 o'clock. Cooled off at 3 hours

26 min, 2 ½ basins *very fine yeast*, fresh stock from brewery, and 6 gal boiling swill. This mash produced 34 *a* 36 gal singlings per charge.

Wednesday, Dec. 14—Therm. 7 AM 86°—7 hhds. At 9 hours 10 min corn soaked as usual, at 10 hours 15 min 16 gal, boiling water 150°, 8 do. 159°, in 30 min rye and malt as above. Cooled off at 12 hours 30 min, 86 *a* 88° 3 basins fine yeast. At 6 PM therm. 44°.

Thursday, Dec. 15—Therm. 36°—7 hhds. As usual. Scald, 150°. Cooled off to 86°. Yeast very good.

6 do. As usual. Scald, 164°. Cooled off to 90°.

Therm. at 6 PM 46°. All these casks were working well at 7 AM on Friday. No material difference perceptible in the time or manner of their fermentation, or the produce.

Friday and Saturday, Dec. 16 and 17—Therm. 46°—Mashed 13 hhds. as usual.

Tuesday, Dec. 20—Therm. 46°—7 hhds. as usual.

Wednesday, Dec. 21—Therm. 42°—6 hhds. Corn soaked as usual, last heat 156° *a* 158°. Cooled off at 3 PM 84 *a* 86° 2 large basins fine yeast and 2 buckets hot swill.

Thursday, Dec. 22—Therm. 42°—7 hhds. Com as usual, last heat 162°. Cooled off to 86° at 1 hour 30 min 2 basins fine yeast.

Fermentation tolerable next day at 7 AM.

Friday, Dec. 23—Therm. 46°—7 hhds. As usual, last heat 158°. Yeast very fine.

Saturday, Dec. 24—Therm. 32°—Mashed 13 hhds.

Monday, Dec. 26—Therm. 32°—7 hhds. Corn soaked as usual stood 2 hours, then 16 gal boiling water 154° *a* 160°, stood 20 min, rye and malt as usual. Cooled off in 2 hours, 84° a 88° 2 large basins good yeast, 38° therm.

Tuesday, Dec. 27—Therm. 36°—7 hhds. Corn soaked as usual stood 2 hours, then 16 gal boiling water 150°, stood 45 min, then 12 gal boiling water 158°, stood 15 min, then rye and malt as usual. Cooled off in 2 hours. Yield 32 gal per charge.

Wednesday, Dec. 28—Therm. 42°—7 hhds. mashed.

Thursday, Dec. 29—Therm. 42°—12 hhds. mashed.

Friday, Dec. 30—Therm. 42°—12 hhds. 6 gal cold and 8 boiling water, 60 lbs corn, stood 1 hour 45 min, then 16 gal boiling then 10 gal boiling water 151 *a* 162°, stood 30 *a* 40 min, then rye and malt as usual. Cooled off to 90°. Therm. at 12 PM 48 *a* 52°.

N. B. The hogsheads used this day were old lime hogsheads, the fermentation better than any preceding. Yield 36 *a* 38 gal singlings per charge.

Saturday, Dec 31—Therm. 47°—12 hhds. As yesterday. Cooled 6 hhds. off to 90° and 6 hhds. to 86°. Fine yeast. Therm. from 6 to 12 PM 55 *a* 52°.

Monday, Jan. 2, 1815—Therm. 50°—12 hhds. As usual. Fine yeast, and good fermentation.

Tuesday, Jan. 3—Therm. 45°—12 hhds. as usual.

Wednesday, Jan. 4—Therm. 45°—12 hhds. As usual. Cooled off to 90°. Fermentation very good. Therm. 12 PM 55°.

Thursday, Jan. 5—Therm. 50°—12 hhds. Corn soaked as usual, stood nearly 2 hours, then 16 gal boiling water, then 8 or 10 gal boiling water 158 *a* 162°, stood 30 *a* 40 min, then rye and malt as usual. Fermentation commenced very fine in 6 hours after cooling off. Therm. at 12 PM 65°.

Friday and Saturday, Jan. 6 and 7—Therm. 50°— 22 hhds. mashed.

Monday, Jan. 9—Therm. 50°—6 hhds. As usual. Cooled off at noon precisely to 80°, fermentation very good at 9 PM.

6 hhds. As usual. Cooled off at 7 PM to 90°. Fermentation very good at 5 AM. Therm. 52°.

Tuesday, Jan. 10—Therm. 50°—12 hhds. As usual.

Wednesday and Thursday, Jan. 11 and 12—Therm. 45°—24 hhds. mashed as usual.

Friday, Jan. 13—Therm. 45°—6 hhds. As usual. Scald 156°. Cooled off to 86°.

6 hhds. As usual. Scald 154. Cooled to 84°. Head on the first 6 hhds. in 6 hours. Therm. 10 PM 50°.

Saturday, Jan. 14—Therm. 49°—Mashed 6 hhds., 6 gal cold, 8 boiling water, 60 lbs corn, stood 1 and a half hours, then 28 gal boiling water 154°, then 25 lbs rye and malt immediately.

N. B. 2 hhds. stood half an hour longer than the others, added 4 buckets boiling water, and cooled off to 98°. Fermentation of the whole commenced at 10 PM *all alike.*

N. B. Doubling this day yielded 38 gal of proof spirit, being the produce of nearly 480 lbs corn and 200 lbs rye and malt. Almost 3 ½ gal per bushel.

Recapitulation.

Date. (1814)		No. of hogsheads mashed	Corn in each hhd	Rye and Malt in each hhd	No. of hhds distilled	Gallons of Spirit	Thermometer heat at 7 AM*
December	5	13	60	30	—	—	40°
	6	6	—	—	—	—	34
	7	13	—	—	—	—	36
	8	12	—	—	—	—	34
	9	7	—	—	—	—	36
	10	6	—	—	4	—	32
	12	7	—	—	10	—	—
	13	14	—	—	4	31	40
	14	7	—	—	6	31	36
	15	7	—	—	6	30	—
	16	—	—	—	14	33	46
	17	13	—	—	4	20	—
	19	—	—	—	6	26	—
	20	7	—	—	6	30	—
	21	6	—	—	6	—	42
	22	7	—	—	8	36	—
	23	7	—	—	12	45	46
	24	13	—	—	12	30	32
	26	7	—	—	8	30	—
	27	7	—	—	8	45	36
	28	7	—	—	8	34	42
	29	12	—	—	8	36	—
	30	12	—	—	8	36	—
	31	12	—	—	12	33	47
January	2	12	—	—	10	33	50
	3	12	—	—	10	33	45
	4	12	—	—	8	33	45
	5	12	—	—	10	34	50
	6	12	—	—	10	36	—
	7	12	—	—	18	34	—
	9	6	—	—	8	54	—
	10	12	—	—	8	33	—
	11	12	—	—	10	48	45
	12	12	—	—	8	32	—
	13	12	—	—	10	16	—
	14	6	—	—	16	18	49
	16	12	—	—	10	51	
	17	12	—	—	—	51	
					286	1,032	Gal 15% over pr'f
					143	155	
				Bushels of Grain	429	1,187	2¾ gal per bushel

*The temperature at 12 AM may be seen by recurring to the days on which it was noted.

Chapter XIII.

Containing observations on Grain.

The preceding directions for mashing are confined to rye and corn, they being most generally used by distillers. Every kind of grain, however, may be made to produce spirit, and as each may occasionally be within the reach of the distiller, he should be acquainted with the proper mode of treatment. Such observations therefore, are here given concerning each as are considered necessary, without interfering with the particular directions for mashing.

I have ever considered the union of rye and corn in mashing, as productive of more spirit and of a purer quality, than can be obtained from either grain alone; and if the proportion of one-fourth part of rye can be obtained, it is enough.

Of Indian Corn.

I believe corn to contain more spirit than any other grain; being of the same weight per bushel, and having little or no bran. Hence it cannot be well fermented alone, as it sinks to the bottom of the cask. Some other grain therefore becomes necessary to enable it to be completely acted upon by the fermentable principle.*

Of rye enough has been said, and it may be sufficient to remark

Of Wheat.

That this grain is of too high price for the distiller, yet it yields more spirit than rye, in the proportion of six to five. It should be mashed nearly at the same heat, or rather lower, and in the same manner.

During the embargo I purchased damaged wheat flour at three dollars a barrel, and mashed it with corn; it answered very well. The heat for mashing appeared to be best between 137 and 147°. Sprouted wheat does well.

*Directions for malting Indian corn are given in a small work on brewing, published at New York about two years since.

On the distillation of Oats.

A mixture of two-thirds corn and one-third oats, will yield upon distillation, a very fine spirit. The oats impart a peculiar flavour, which I think preferable to that given by rye; there is however so large a quantity of bran in, or hulls upon oats, that it is difficult and even dangerous to work it; unless it has undergone a very complete fermentation it will rise in the still, and the hull being light, is thrown up and adheres to the head, or if the fire is very strong the whole is forced into the worm which is very soon choaked and endangers the bursting of the head. In such case the fire should be immediately extinguished and the plug in the breast of the still knocked out.

Sixty pounds of oats will produce three gallons.

The directions for mashing rye with corn should be followed in mashing oats, except that oats do not require to be scalded quite so high as rye.

Could any method be employed by which to separate the hull, it would be attended with less trouble both in mashing and distillation. I was always however of the opinion that the *head* formed by the oats on the top of the hogsheads, was very favourable to fermentation on account of the complete exclusion of the external air; but I frequently took off this head or top, which consisted almost entirely of bran, before putting the wash into the still, when this was done, there was no trouble in distillation.

Of Buckwheat.

Buckwheat is very serviceable to mix with corn when a sufficient quantity of rye cannot be had; it does not however answer to work alone, but with corn it may be worked very nearly according to the directions given for rye; it must not be scalded quite so high as rye, and as it is disposed to a very rapid fermentation, care must be taken to *cool off* lower, and not quite so much yeast should be used as with rye: a lesson may be had upon this subject from the making of buckwheat cakes where they have good yeast. Buckwheat weighs about 33 pounds to the bushel, and will yield three gallons to 60 pounds.

Cockle,

When accidentally mixed with other grain excites a very rapid fermentation if in any quantity, and it renders the spirit very fiery. It certainly produces a great quantity of *spirit*, and is not of material disadvantage to the *yield*.

Barley

Is a very hard flinty grain and does not work well unless malted. It is not much raised by fanners except in the neighbourhood of breweries, where the price is generally so high that it is not an object to the distiller, except for malt, and even for this, rye will be mostly found cheaper. It should be mashed like corn.

Garlic,

If mixed in any considerable quantity with rye, will hinder it from grinding well, and consequently lessen the product. It imparts a disagreeable flavour to the spirit, and abounds most in grain of *inferior quality*, which is another reason why a bushel of pure rye, will produce more than the same measure of garlicky rye, for *garlic does not yield any spirit.*

The weight and produce of different kinds of Grain will be found to be nearly as follows.

Grain	Amount (lbs)	Yield (qts per bushel)
Wheat	60	12 to 16
Rye	60	12 to 16
Corn	60	12 to 16
Buckwheat	33	5 to 6
Oats	32	6 to 8
Barley	45	7 to 9
Rice*	70	14 to 16

*The author has tasted arrac upwards of fifteen years old, said to have been made from rice. It was remarkably strong, and finely flavoured, partaking of both Cogniac brandy and old cane juice. Of the process, and precise produce per bushel, he is ignorant.

Chapter XIV.

Of Malt.

The celebrated Irish whiskey is made principally from malted grain. Attempts have been made to imitate it in this country, but have been generally unsuccessful. How far this is practicable, and whether if so, it would be more profitable than our present method of distilling, requires consideration. To the speculative it may afford matter of conjecture, why the making of Irish whiskey is given up as wholly impracticable, although attended by no *secret* art or hocus pocus, by the very men who are constantly puzzling their brains about Holland gin, and the secret by which the Hollanders are enabled to excel us in that article. Is it not probable that the same cause operates in both cases? If not, why is not Irish whiskey (about which no secret is pretended) made in this country, equal to what is imported?

Not having a malt kiln attached to my distillery, and it being generally difficult for me to obtain malt, I have made no experiments on the subject. There are many objections to the general introduction of malt distilleries. These however should not operate against individual attempts to imitate Irish whiskey, as the making of two gallons of it from a bushel of malt would be the most productive kind of distilling, so long as the present high price of that article continues. But as there can scarcely be said to be a proportionate medium between the price of our common whiskey and Irish whiskey, a malt distillery would certainly not answer unless capable of producing a perfect imitation of Irish whiskey.

In most of the Atlantic states, and all places within the influence of breweries, the high price of barley places it beyond the reach of the distiller. But in the western country, barley may be raised for such price as to render it an object to the distiller, and hence we may expect to derive, at no very distant day, our supplies of malt spirit. So long however as Indian corn continues to be a staple of our country, it will constitute a principal source for our *national beverage*, and other grains malted, will be used as a mean of facilitating our operations and increasing their product.

Malt then may be made of rye or barley, according to the price or facility of procuring either grain, it being now pretty generally allowed, that any advantage which one may have over the other, depends upon these circumstances. There are a variety of opinions as to the proper method of making malt for a distillery, not only as to the extent of the germination, but to the drying afterwards. These however will generally be found to be *distinctions without a difference*, and probably owing to the fact, that the same care is not necessary in preparing malt for the distillery as for the brewery; a circumstance not attended to in the attempts at malting in different distilleries. It is a common practice in small distilleries to make a bushel or two of malt at a time, and dry it on the stills, and this is preferred to malt made by the brewer. I have frequently used malt of this kind, and never observed any difference in their results. Yet it is very evident, that malt made in this way must be imperfect, and far from possessing the virtues ascribed by maltsters to well made malt. This subject therefore, would seem worthy of investigation.

It has been the opinion of some distillers that there is a particular quantity of malt *necessary*, than which neither more nor less could be used without loss. And of others that the greater the quantity of malt, even were all the grain malted, the better. On the contrary, it has been discovered by the Scotch and English distillers, that "*by mixing good grain reduced to meal, with their malt, that they obtain more spirit than from an equal quantity of good malt*."* Grain also which has sprouted in the field, and whose vegetation has been stopped, and thence concluded not proper for vinous fermentation, the change into saccharine matter not being perfect, upon being mixed with a quantity of malt and fermented, was found to furnish as much spirit as if the whole had been in the state of perfect malt. Without pretending to reconcile, or account for, these different opinions and facts, a mention of which was considered as proper, I will merely observe that the most proper quantity appeared to me from different experiments, to be from ten to fifteen pounds for each hogshead, in cold weather, and in warm weather a smaller quantity may do. The Holland distillers use about twenty pounds of malt and seventy pounds of raw grain.

To the mere practical distiller, it may be sufficient to know how to make the most of his materials; to him the operation of causes and effects is not of primary importance. But to the philosopher, and the man of science, whose business and whose pleasure it is, to search out hidden causes, and unfold the secrets of nature, here is a subject not unworthy of attention. If the vegetation which the grain undergoes in the process of malting, forms the saccharine matter, and if it be considered that this saccharine matter, alone and exclusively, produces the spirit, then, doubtless malted grain should yield more spirit, and of a more pleasant taste, than raw grain, in proportion as the process may be perfected. But the experience and practice of all distillers appear to prove this to be a mistaken theory; or have they made their experiments, or continued their practice, without sufficient attention to those circumstances and precautions actually necessary to obtain a correct result? I rather think not.

What matter then is added to or disengaged from grain by the process of malting, or what circumstance so changes its nature, that by the mixture of a certain proportion of raw grain with it, more spirit is produced than can be obtained from an equal quantity of malt or an equal weight of grain unmalted?

Queries of this nature might be multiplied without end; I leave the matter however for the investigation of more able heads, and will proceed with matter more immediately within the design of the present work.

*See *Irwin's Chemical Essays*, page 318.

The distiller who is near a malt kiln, where wood is high, will generally find it to his advantage, to purchase malt, it frequently happening that a kiln of malt may be dried too much for brewing, which will yet answer for distilling, and consequently may be obtained rather below the market price. If however, he is on a sufficiently large scale to use eight or ten bushels of malt per day, without exceeding the proportion mentioned in the preceding directions, it will be better for him to build a kiln, and make his own malt. Or, if he be situated where malt must be hauled from a distance, he must make for himself. He may in either case then, have the great advantage, of having the full proportion of malt necessary, at a trifling addition of expense to the price of the raw grain, and very little risk or trouble.

He who undertakes to make malt will find the process simple, though requiring very close attention where perfection is expected, and for this end should have a complete maltster; but for the generality of distillers, who do not use a large quantity, the directions in the ensuing chapter will suffice.

To the preceding remarks upon malt which will probably be found sufficient for practical men, and distillers, it may not be improper to add a few observations on "the nature and properties of malt" from "Theoretic hints on an improved practice of brewing malt liquors," by Mr. Richardson.

However we may differ in our opinions on the necessity of perfecting the process of malting for our distilleries, the remarks of Mr. R. may be of service to those who attempt to make malt.

"The process of making malt is an artificial or forced vegetation, in which the nearer we approach the footsteps of nature in her ordinary progress, the more certainly shall we arrive at that perfection of which the subject is capable. The farmer prefers a dry season to sow his corn in, that the common moisture of the earth may but gently insinuate itself into the pores of the grain, and thence gradually dispose it for the reception of the future shower, and the action of vegetation. The maltster cannot proceed by such slow degrees, but makes an immersion in water a substitute for the moisture of the earth, where a few hours infusion is equal to many days employed in the ordinary course of vegetation; and the corn is accordingly removed as soon as it appears fully saturated, lest a solution, and consequently a destruction of some of its parts, should be the effect of a longer continuance in water, instead of that separation which is begun by this introduction of aqueous particles into the body of the grain.

"Were it to be spread thin after this removal, it would become dry, and no vegetation would ensue: but being thrown into the couch, a kind of vegetative fermentation commences, which generates heat, and produces the first appearance of germination. This state of barley is nearly the same with that of many days continuance in the earth after sowing: but being in so large a body, it requires occasionally to be turned over, and spread thinner; the former to give the outward parts of the heap their share of the required warmth and moisture, both of which are lessened by exposure to the air; the latter to prevent the progress of the vegetative to the putrefactive fermentation, which would be the consequence of suffering it to proceed beyond a certain degree.

"To supply the moisture thus continually decreasing by evaporation and consumption, an occasional but sparing sprinkling of water should be given to the floor, to recruit the languishing powers of vegetation, and imitate the shower upon the corn field. But this should not be too often repeated; for, as in the field, too much rain, and too little sun, produce rank stems and thin ears, so here would too much water, and of course too little dry warmth, accelerate the growth of the malt, so as to occasion the extraction and loss of such of its valuable parts, as by a slower process would have been duly separated and left behind.

"By the slow mode of conducting vegetation here recommended, an actual and minute separation of the parts takes place. The germination of the radicles and acrospire carries off the cohesive properties of the barley, thereby contributing to the preparation of the saccharine matter, which it has no tendency to extract or otherwise injure, but to increase and meliorate, so long as the acrospires is confined within the husk; and by how much it is wanting of the end of the grain, by so much does the malt fall short of perfection, and in proportion as it has advanced beyond, is that purpose defeated.

"This is very evident to the most common observation, on examining a kernel of malt in the different stages of its progress. When the acrospire has shot but half the length of the grain, the lower part only is converted into that yellow saccharine flower we are solicitous about, whilst the other half affords no other signs of it than the whole kernel did at its first germination. Let it advance to two thirds of the length, and the lower end will not only have increased its saccharine flavour, but will have proportionally extended its bulk, so as to have left only a third part unmalted. This, or even less than this, is contended for by many maltsters, as a sufficient advance of the acrospire, which they say has done its business as soon as it has passed the middle of the kernel. But we need seek no further for their conviction of error, than the examination here alluded to.

"Let the kernel be slit down the middle, and tasted at either end, whilst green, or let the effects of mastication be tried when it is dried off; when the former will be found to exhibit the appearances just mentioned, the latter to discover the unwrought parts of the grain, in a body of stony hardness, which has no other effect in the mash tub than that of imbibing a large portion of the liquor, and contributing to the retention of those saccharine parts of the malt which are in contact with it; whence it is a rational inference, that three bushels of malt, imperfect in this proportion, are but equal to two of that which is carried to its utmost perfection. By this is meant the farthest advance of the acrospire, when it is just bursting from its confinement, before it has effected its enlargement. The kernel is then uniform in its internal appearance, and of a rich sweetness in flavour, equal to any thing we can conceive obtainable from imperfect vegetation. If the acrospire be suffered to proceed, the mealy substance melts into a liquid sweet, which soon passes into the blade, and leaves the husk entirely exhausted.

"The sweet thus produced by the infant efforts of vegetation, and lost by its more powerful action, revives and makes a second appearance in the stem, but is then too much dispersed and altered in its form to answer any of the known purposes of art.

"Were we to inquire by what means the same barley, with the same treatment, produces unequal portions of the saccharine matter in different situations, we should perhaps find it principally owing to the different qualities of the water used in malting. Hard water is very unfit for every purpose of vegetation, and soft will vary its effects according to the predominating qualities of its impregnations. Pure elementary water is in itself supposed to be only the vehicle of the nutriment of plants, entering at the capillary tubes of the roots, rising into the body, and there dispersing its acquired virtues, perspiring by innumerable fine pores at the surface, and thence evaporating by the purest distillation into the open atmosphere, where it begins anew its round of collecting fresh properties, in order to its preparation for fresh service.

"This theory leads us to the consideration of an attempt to increase the natural quantity of the saccharum of malt by adventitious means; but it must be observed on this occasion, that no addition to water will rise into the vessels of plants, but such as will pass the filter; the pores of which appearing somewhat similar to the fine strainers or absorbing vessels employed by nature in her nicer operations, we by analogy conclude, that properties so intimately blended with water as to pass the one, will enter and unite with the economy of the other, and *vice versa*.

"Supposing the malt to have obtained its utmost perfection, according to the criterion here inculcated, to prevent its further progress and secure it in that state, we are to call in the assistance of a heat sufficient to destroy the action of vegetation, by evaporating every particle of water, and thence leaving it in a state of preservation, fit for the present or future purposes of the brewer.

"Thus having all its moisture extracted, and being by the previous process deprived of its cohesive property, the body of the grain is left a mere lump of flour so easily divisible, that, the husk being taken off, a mark may be made with the kernel as with a piece of soft chalk. The extractable qualities of this flour are, a saccharum closely united with a large quantity of the farinaceous mucilage peculiar to bread corn, and a small portion of oil enveloped by a fine earthy substance, the whole readily yielding to the impression of water applied at different times, and different degrees of heat, and each part predominating in proportion to the time and manner of application.

"In the curing of malt, as nothing more is requisite than a total extrication of every aqueous particle, if we had in the season proper for malting, a solar heat sufficient to produce a perfect dryness, it were practicable to produce beer nearly colourless; but that being wanting, and the force of custom having made it necessary to give our beers various tinctures and qualities resulting from fire, for the accommodation of various tastes, we are necessitated to apply such heats in the drying as shall not only answer the purpose of preservation, but give the complexion and property required."

Chapter XV.

To make Malt.

Steep the rye or barley in water, until it can be nearly mashed endwise between the fingers: this, in warm weather, will require from eighteen to twenty-four hours, and longer in cold weather. Then drain off the water, and throw the malt into a heap, on an earthen floor, if possible, until it has begun to sprout, which will be in about eighteen hours. In cold weather it will sometimes be necessary to cover it with a blanket and sprinkle it with warm water, in order to accelerate the sprouting; or make it *come*, as it is termed. It is now to be spread out to about the depth of six inches, and turned occasionally, that it may all come alike. When the sprout is as long as the grain, and before the blade and spire begins to put out, it must be spread very thin upon a dry floor, in order to put a stop to any further growth, and in thirty-six or forty-eight hours it will be fit for the kiln, where it must be made perfectly dry.

Chapter XVI.

To dry Malt.

By a little attention, the surplus heat from the stills may be employed, and will be sufficient to dry malt. For this purpose, in building the chimney of the distillery, leave a hole about six inches square in one side of the chimney, a small distance above the second floor; and another of the same size on the other side, a little higher up: then on the floor, lay a coat of clay about four inches thick; to prevent the floor from taking fire; and four and a half feet long and three and a half feet wide. Build around this a brick wall sixteen inches high, and from the chimney to within six inches of the end wall, run a thin wall, of the same height. Cover the whole with a piece of sheet iron. Then by placing a damper in the chimney, between the two holes, the heat will be thrown into the kiln, pass round the intermediate wall and must be conducted to the upper hole by a small flue.

In this way four or five bushels a week may be dried.

Drying Malt by Steam.

A patent has been obtained by James Adam, esq. for the purpose—the method consists in the application of heat from steam, which may be most conveniently done by confining the steam within a chest or chests, or in hollow cylinders, or other vessels, of any form or shape suitable to the purpose, on a floor of metal, pottery, or other substance or substances, which most easily transmit heat, and which being formed steam-tight, permit the heat to pass through the same without any steam or moisture; and that the malt and other grain being spread upon the floor is thereby dried in an equable, gentle, and regular manner, and the degrees of heat may be easily regulated by the admission of more or less steam, so that considerable precision in the degree of heat given to all parts of the floor may always be attainable, and the malt or grain stirred in the usual way.

For a more particular description see the 21st volume *Repertory of Arts.*

Of Grinding.

As the grinding of the grain is the business of the miller, it is too frequently left to his discretion to judge of the coarseness or fineness which may be most proper. This however is a matter of which the distiller is the only proper judge, and is sufficiently important to require his particular attention. On this subject as on many others, distillers differ very much, some requiring rye to be ground very coarse, others very fine, while others go so far as to say it should only be *ground upon country stones and very slowly!*—As to this I cannot judge; it is true, that rye is sometimes so ground as to render it dead or inert and difficult to ferment, but this rarely happens with a tolerably good miller. With respect to the degree of fineness there is a medium, which may be best discovered by the observation and experience of the distiller himself. It should neither be so fine as is requisite for boulting, nor so coarse as is generally chopt for horse feed. In the latter case it is subject to a considerable loss, inasmuch as it resists the impression of the hot water in scalding. It will not form a proper union with the water, consequently cannot be made to ferment perfectly, it however imbibes a sufficient quantity of moisture to make it turn sour; the acidity is communicated to the whole cask, and it tends rapidly to an acetous instead of a vinous product.

When ground very fine it is also subject to disadvantages, though not so great as above mentioned. The greatest danger arises from overscalding, which renders it clammy, and apt to adhere to the sides and bottom of the still.

A medium therefore is to be observed, which will be best ascertained by experiment.

Indian corn, from the flintiness of its nature, is difficult to dissolve, it therefore cannot be ground too fine.

Oats and buckwheat should be ground about as fine as rye.

The grinding of malt is not of much importance; I never found any material difference in the result whether it was merely chopped or ground very fine.

Barley when used without malting should be ground very fine.

Chapter XVII.

Of Vinous Fermentation.

There are three species of fermentation, the vinous, the acetous, and the putrefactive, each being distinguished by its products, as well as by the phenomena it presents. It has been supposed that these three succeed each other in invariable order, that the vinous always precedes the acetous, and that this equally precedes the putrefactive. Some facts are in favour of this hypothesis, or, there are substances which undergo these successive changes.

Many weak vinous liquors, by a continuance of the fermentative process, become sour, forming vinegar; and vinegar also undergoes decomposition, forms a mould, or passes into a species of putrefaction. But it is not to be concluded that these kinds of fermentation invariably succeed each other; many vegetable substances become sour, which we do not discover ever to assume any vinous state, and a still greater number undergo that decomposition analogous to putrefaction, without having passed through the other two stages of fermentation. When they do succeed one another, however, the vinous is that which precedes the others; and it never succeeds them.

This important process by which saccharine solutions are converted into intoxicating liquors is one of the most complicated in chemistry, and the precise cause of this change is as yet imperfectly known. It is therefore merely intended here to notice the conditions requisite to fermentation, the appearances that occur during the process, and the essential product of it.

Of the vegetable principles, saccharine matter is that which passes with most facility and certainty into the vinous fermentation; and fermented liquors are more or less strong, as the juices from which they have been formed have contained a greater or less proportion of sugar before fermentation, for the addition of sugar to the weakly fermentable juices will enable them to produce a strong full bodied liquor, and the most essential exit in this process is the disappearance of the sugar, and the consequent production of alcohol.

Certain circumstances, however, are necessary to enable it to commence and proceed. These are, a due degree of dilution, in water, a certain temperature, and the presence of substances which appear necessary to favour the subversion of the balance of affinities by which the principles of the saccharine matter would otherwise be retained in union, or at least would be prevented from entering into those combinations necessary to form vinous spirit. These substances, from this operation, are named ferments.

1. A certain proportion of water to the matter susceptible of fermentation is requisite. If the latter is in large quantity proportioned to the water, the fermentation does not commence easily, or proceed so quickly; on the other hand, too large a proportion of water is injurious, as causing the fermented liquor to pass speedily into the acetous fermentation. The necessary consistence exists naturally in the juice of grapes, and in the saccharine sap of many trees; and other spontaneously fermentable liquors: for, if these very liquors be deprived by gentle evaporation of a considerable portion of their water, the residue will not ferment until the requisite consistence is restored by the addition of a fresh portion of water.

2. A certain temperature is not less essential; it requires to be at least 55° of Fahrenheit. At a temperature lower than this, fermentation scarcely commences, or if it has begun, proceeds very slowly; and if too high, requires to be checked to prevent it from passing into the acetous state. Lastly, though sugar or substances analogous to it are the matters which serve as the basis of fermentation, and from which its products are formed, the presence of other matter is requisite to the process; it has been often stated indeed, that sugar alone, dissolved in a certain quantity of water, and placed in a certain temperature, will pass into a state of fermentation. It is doubtful however, if this happens with a solution of pure sugar; and any change which is observed is imperfect and irregular; nor does the liquor become vinous, but rather sour.

The substance usually added to produce fermentation is called yeast: but Mr. Cooper says, "by ferments, we mean any substance which being added to any rightly disposed fermentable liquor, will cause it to ferment much sooner and faster than it would of itself, and consequently render the operation shorter, in contradiction to those abusively called so, which only correct some fault in the liquor, or give it some flavour."

When the proper sort of ferment is pitched upon, the operator is next to consider its quantity, quality, and manner of application. The quantity must be proportioned to that of the liquor, to its tenacity, and the degree of flavour it is intended to give, and to the despatch required in the operation. From these considerations he will be able to form a rule to himself; in order to the forming of which, a proper trial will be necessary to shew how much suffices for the purpose.

The greatest circumspection and care are necessary in regard to the quality of the ferment, if a pure and well flavoured spirit be required. It must be chosen perfectly sweet and fresh, for all ferments are liable to grow musty and corrupt; and if in this state they are mixed with the fermentable liquor, they will communicate their nauseous and filthy flavour to the spirit, which will scarcely ever be got off by any subsequent process. If the ferment be sour, it must by no means be used with any liquor, for it will communicate its flavour to the whole, and even prevent its rising to a head, and give it an acetous instead of a vinous tendency.

When the proper quantity of a well conditioned ferment is prepared it should be diffused in the liquor to be fermented, in a tepid or lukewarm state; when the whole is thus set to work, secured in a proper degree of warmth, and kept from a too free intercourse with the external air, it becomes as it were the sole business of nature to finish the operation and render the liquor fit for the still.

The first signs of fermentation are, a gentle intestine motion, the rising of small bubbles to the top of the liquor, and a whitish turbid appearance. This is soon followed by the collection of a froth or head, consisting of a multitude of air bubbles entangled in the liquor, which as the process advances, rise slowly to considerable height, forming a white dense permanent froth. A very large portion of the gas also escapes, which has a strong, penetrating, agreeable, vinous odour. The temperature of the liquor at the same time increases several degrees, and continues so during the whole of the process. Sooner or later these appearances gradually subside, the head of the foam settles down and the liquor appears much clearer and nearly at rest, having deposited a copious sediment, and from being viscid and saccharine, is now become vinous, intoxicating, much thinner, or of less specific-gravity.

The process of fermentation however does not terminate suddenly, but goes off more or less gradually according to the heat at which it was commenced, and of the temperature of the external air.

The gas of fermenting liquors has been long known to consist for the most part of carbonic acid; it will therefore extinguish a candle, destroy animal life, convert caustic alkalies into alkaline carbonats, and render lime water turbid by recomposing lime stone, which is insoluble, from the quick lime held in solution.*

But beside the carbonic acid, it has been proved by Scheele, to hold in solution a sensible quantity of alcohol, and Proust has detected in it a portion of azot. Mr. Collier (*Manchest. Trans.* Vol. 5) has further shown, that in this gas, are contained all the requisites for vinous fermentation. He passed the whole of the gas from a nintey gallon fermenting tub into a cask of water, and divided the liquor thus impregnated into three parts, of which one being immediately distilled, afforded a small quantity of alcohol; to the second was added some yeast, by which a new fermentation was excited, and the subsequent product of distilled spirit was nearly doubled, and the third being suffered to ferment a longer time, produced vinegar.

The attenuation of liquors, or the diminution of their specific gravity by fermentation, is very striking. This is shown by the hydrometer, which swims much deeper in fermented liquor than in the same materials before fermentation.

Much of this attenuation is, doubtless, owing to the destruction of the sugar, which dissolves in water, adds to its density, and to the consequent production of alcohol, which on the contrary by mixture with water diminishes the density of the compound.

The tract or mucilage also appears to be in some degree destroyed by fermentation, for the gelatinous consistence of thick liquors is much lessened by this process: the destruction of this principle, however, is by no means so complete as of the sugar; many of the full-bodied ales, for example, retaining much of their clamminess and gelatinous density even after having undergone a very perfect fermentation.†

*If a clear saturated solution of pearl ash, be exposed in an open vessel, within a vat of fermenting liquor, beautiful chrystals of salt, formed of the alkali and acid or air, will be formed. From these the air may be again disengaged by vinegar or any other acid. These chrystals being thrown into a barrel of stale cyder, at the rate of not more than four ounces to a barrel, and the bung being replaced quickly will be found to diminish the acidity, and increase its briskness without injuring its salubrity.

†This arises from their having more extract than there is water to decompose. The same paucity of water produces the same effects in the fermentation of syrup.

It has been doubted whether alcohol be the product of vinous fermentation, or of the subsequent distillation, by which it has always been obtained, or in other words, whether alcohol exists ready formed in the fermented liquor, or is the product of new combinations resulting from the subsequent application of heat in distillation. The proofs which have been brought forward in opposition to theory, are chiefly founded on the researches of Fabroni,* who attempted to separate alcohol by saturating the wine with dry sub-carbonate of pot-ash, but did not succeed, although by the same means he could detect very minute portions of alcohol which had been purposely added to the wine.

In favour of this opinion, that alcohol is formed by the process of fermentation, we have the experiments of Wm. Thomas Brande, Esq.†; and although we agree with him in opinion, we do not think his experiments decisive on the subject. Mr. Brande added one seventh part of alcohol to wine, and could not separate it by means of sub-carbonate of pot-ash, though Fabroni says, that by the same means he could separate one hundredth part when mixed with wine. But when Mr. B. added one fourth of alcohol to wine, the sub-carbonate of pot-ash would separate it in a very impure state.

In another experiment Mr. B. added four ounces of dry and warm sub-carbonate of pot-ash to eight-fluid ounces of port wine, which was previously ascertained to afford by distillation 20%, of alcohol (by measure) of the specific gravity of 0,825 at 60°. In twenty-four hours the mixture had separated into two distinct portions; at the bottom of the vessel was a strong solution of the sub-carbonate, upon which floated a gelatinous substance, of such consistency, as to prevent the escape of the liquor beneath when the vessel was inverted, and which appeared to contain the alcohol of the wine, with the principal part of the extract tan and colouring matter, some of the sub-carbonate and a portion of water; but as these experiments relate chiefly to the spirit contained in the wine, the other ingredients were not minutely examined.

To seven fluid ounces of the same wine, were added one fluid ounce of alcohol of 825, and the same quantity of the sub-carbonate as in the last experiment; but after twenty-four hours had elapsed, no distinct separation of alcohol had taken place.

If the spirit afforded by the distillation of wine was a product, and not an educt, Mr. Brande conceived, that by performing the distillation at different temperatures, different proportions of spirit would be obtained.

The following are the experiments made to ascertain this point. Four ounces of dried muriate of lime were dissolved in eight fluid ounces of port wine, employed in the former experiments by this addition: the boiling point of the wine which was 190° Fahrenheit, was raised to 200°; the solution was put into a retort placed in a sand heat, and was kept boiling until four fluid ounces had passed over into the receiver, the specific gravity of which was 0,96316 at 60° Fahrenheit.

The experiment was repeated with eight fluid ounces of the wine without any addition, and the same quantity was distilled over as in the last experiment, its specific

**Annales de Chimie*, 31st vol. p. 303.

†See *Repertory of Arts*, vol. 20, p. 144. New Series.

gravity at 60° was 0,96311. Eight fluid ounces of the wine were distilled in a water bath; when four fluid ounces had passed over the heat was withdrawn. The specific gravity of the liquor in the receiver was 0,96320 at 60° Fahrenheit.

The same quantity was distilled at a temperature not exceeding 180°. This temperature was kept up from four to five hours for five successive days, at the end of which period, four ounces having passed into the receiver, its specific gravity at 60° was ascertained to be 0,96314.

It may be concluded, from these results, that the proportion of alcohol is not influenced by the temperature at which wine is distilled; the variation of the specific gravities in the above experiments being even less than might have been expected, when the delicacy of the operation by which they are ascertained is considered.

Mr. B. says, he repeatedly endeavoured to separate the spirit from wine, by subjecting it to low temperatures, with a view to freeze the aqueous parts; but when the temperature is sufficiently reduced, the whole of the wine forms a spongy cake of ice.

After detailing all these experiments, we must say that they do not yield any positive conclusion on the subject.

The first experiment which comes nearest to proving the point is unsatisfactory. He says that the gelatinous substance which floated on the surface appeared to contain the alcohol, but he does not say what appearance it was that induced him to conclude that the spirit was in it. If he had thrown it on the still head and applied a blaze to it, and had found that it burnt much more strongly than an equal quantity of the wine from which it came; and if the liquid that held the sub-carbonate in solution had yielded no inflammable vapour when thrown on the still head, he would have come nearer to certainty on the subject.—But even this would not have satisfied his opponents. They require the separation of the spirit without changing it into vapour, because they allege, that it is produced either by the application of the heat, or when the fermented liquor is in a state of vapour. If the spirit had been contained in this gelatinous substance, we would suppose, that by means of some of the various acids and alkalies or of strong alcohol, an agent might readily have been found capable of separating it. The latter experiments only show that spirit may be obtained in distilling with a low as well as high heat.

Our opinion that the spirit is formed by the process of fermentation, and not in the distillation of the vinous liquid, arises from observing the same intoxicating effects on the human body, from drinking fermented liquors, as from drinking the spirit distilled from them, the degree of intoxication being proportional to the quantity of spirit which the fermented liquors will yield on distillation, and because we hold it unphilosophical to conclude, until it has been proved by experiment, that the mere volatilization of this fluid changes its nature, contrary to what happens in volatilizing every other fluid that we know.

Hence we conclude that the production of alcohol is one of the last efforts, or the completion of the process of fermentation.

The atmospheric air seems to have no share whatever in vinous fermentation, for it will take place full as well in closed as in open vessels, provided, space is allowed for the expansion of the materials and the copious production of gas. Indeed Mr. Collier found by direct experiment that more spirit is produced by close, than open fermentation. In three separate experiments, in each of which an equal quantity of wort and yeast were fermented, under circumstances precisely similar, with the single exception, that in one the vessel was open, and in the other closed (the gas having no exit but through a tube dipped in water) he found on distilling each fermented liquor, and drawing off the same quantity of spirit from each, that the liquor from the close vessel was constantly of less specific gravity, and therefore, richer in alcohol than the other. Where the spirit from the open vessel was 24 degrees below proof, that from the close, was 56 degrees; where the former was 83, the latter 65; and where it was 103, the other was 93.

The results of Lavoisier's experiments should not pass unnoticed, though it is obvious that much too great simplicity is attempted in the explanation of a process, which every circumstance shows to be very complicated. The simple points to which the experiments of this able enquirer tend, is (setting aside all other agents) to explain how sugar becomes converted into carbonic acid and alcohol, which after all, is the characteristic phenomenon of vinous fermentation.

The entire products of sugar, yeast, and water, fermented in close vessels, are stated to be carbonic acid, alcohol and water, together with a small portion of acetous acid; and from these facts the following theory is deduced: sugar is composed of eight parts hydrogen, 64 oxygen, and 28 carbon, and the process of fermentation effects a change merely in the arrangement of the constituent parts of the sugar, converting one portion into carbonic acid, and the other into alcohol; and hence (as carbonic acid contains only carbon, with a large proportion of oxygen) the portion which is left must contain all the hydrogen, part of the carbon, and a very small portion of oxygen; or, in other words, by this new arrangement of the ingredients of the sugar, one portion, namely, the carbonic acid, is totally deprived of hydrogen, and overloaded with oxygen, while the other portion, namely, the alcohol, abounds in hydrogen, and is deficient in oxygen, the carbon being divided between the new products in nearly an equal proportion with regard to their respective qualities.*

Nothing more plausible than the above has, perhaps, hitherto been offered to the general phenomena of vinous fermentation, though it is very defective in many essential parts, and even does not correspond with the alleged composition of alcohol, given by the same chemist in another part of his inquiries.

The great question remaining to be determined by future inquirers is, what may be the substance or circumstance, which disposes sugar to ferment; for it has been proved that sugar will not of itself begin this spontaneous change into carbonic acid and alcohol, though when once begun, the process will probably go on without further assistance.

Some of the most common fermenting ingredients as the sweet infusion of malt, technically called wort, it is well known will slowly enter into fermentation without the addition of *yeast*: Hence chemists have sought in this substance for the principle which gives the first impulse to the fermentation of sugar.

*Is it not more probable that the water is decomposed, its oxygen combined with the carbon of the sugar to form fixed air, and its hydrogen with the other constituent parts of the fermentable matter to form alcohol? This is the more likely as the extent of fermentation is exactly regulated by the quantity of water present.

Lavoisier remarked, that, the matter of yeast is a compound of carbon, hydrogen, oxygen and nitrogen, and is therefore so far of an animal nature. From the more recent researches of Fabroni, Thenard and Seguin, it is proved, that it is a substance analogous to gluten or albumen which exists in yeast, which is derived from those vegetable juices or infusions that without any addition are capable of fermenting, and which excites the vinous fermentation.

Fabroni, as the result of his researches on fermentation announced, that though saccharine matter is the principle necessary to vinous fermentation, it does not ferment alone, but only by the aid of another substance capable of acting upon it; this substance is the glutinous or vegeto-animal matter, which exists in the nutritive grains, and which, as he stated, is also contained in the grape, being deposited from its juice. When the deposition is complete, or rather when the glutinous matter is perfectly separated by repeated filtrations, the juice does not ferment, but the sediment, mixed with a substance susceptible of fermentation causes it to pass into that state.

It is this vegeto-animal matter, according to Fabroni, which principally constitutes the yeast of wine or beer: and Rouelle long ago found, that the sediment deposited from wines in fermentation is of this nature, as it affords much ammonia when decomposed.

Mons. Thenard of Paris, found that this substance was deposited from the juice of the grape, the cherry, apple, pear, and other fruits, during their fermentation; and those which afforded the largest quantity of it, were those which run most quickly into the vinous fermentation. It is always deposited too, as their fermentation proceeds; and a sediment of yeast always appears, when alcohol is formed.

When dried, it is still capable of exciting fermentation, and may be preserved in this state for an indefinite time.

These researches however still leave a degree of obscurity with regard to this principle; whatever may be its nature, it must exist in the sweet vegetable juices, and in the infusions of the grains that have been subjected to malting, since these are capable of passing into the vinous fermentation without the addition of yeast, and even deposit it as the process proceeds.

It has been generally supposed, that no substance enters into the vinous fermentation, except sugar, or from, which sugar may be extracted; and that the process of malting grain was necessary to develop the sugar or saccharine matter to render it susceptible of vinous fermentation. The practice however of grain distillers proves this to be a mistake, as they obtain as much spirit from a mixture of malted with unmalted grain, as if the whole were malted.

That fecula alone, or at least mixed with no more saccharine matter than what is contained in the grain, may be made to ferment, is established by the experiments of Fourcroy and Vauquelin.

Twenty-four pounds of the flour of the unmalted barley, having been put into a vat with seven times its weight of warm water, and four pounds of yeast, entered into a strong fermentation, which continued several days. The liquor at the end of that time, submitted to distillation gave a weak spirit, which by rectification afforded alcohol. The quantity amounted to twenty-three ounces.

Now Lavoisier had established, that 100 parts of sugar give 58 of alcohol; and as twenty-four pounds of unmalted barley contained only five ounces of sugar, it follows that four times more alcohol had been formed than that sugar could have furnished; a large quantity of it had therefore been formed from the fecula of the grain.

It is suggested, that even in this process, the fecula may proceed rapidly through the intermediate state of sugar, in passing into the vinous fermentation*; from the circumstance, that when the flour of unmalted grain is mixed with a quantity that has been malted, the whole, mixed with water, becomes sweet. This however, only proves, that the flavour of the sugar is stronger than that of the fecula.

Our knowledge of the series of chemical changes which constitute vinous fermentation, is still imperfect. The facts are not yet sufficiently ascertained to admit of any certain conclusions being drawn, much less of a perfect theory being delivered, since amid the theory which still prevails, various suppositions, *a priori*, perhaps equally probable, might be formed as to the reciprocal actions of substances, the elements of which are so much disposed to mutual combinations, as those concerned in fermentation; and the subject must still remain to be elucidated by farther research.

The properties of the fermented liquor, its odour, pungency, and intoxicating quality are owing to the presence of a substance which can be separated from it by distillation, and which, in a pure state, possesses these qualities in a much higher degree. It constitutes in the state of dilution in which it is obtained by distillation, vinous spirit, or, as obtained from the different fermented liquors, from which it derives peculiarities of taste and flavour, the spirituous liquors of commerce. These, by certain processes, afford this principle pure, and the same from all of them: in this pure state it is called spirits of wine or alcohol.

*There can be no doubt of this fact in the mind of any one who has attended to the process of mashing in our grain distilleries. By the high heat at which the unmalted grain is kept for some time, the same change into saccharine matter takes place as is produced by the process of malting. This is evidenced by the taste, which is perfectly *sweet*, if the proper degree of heat has been observed; but otherwise, little or no sweetness is perceptible. And the produce in spirit is always in proportion to the sweetness of the *mash* or wort previous to fermentation.

Chapter XVIII.

Of Hops.

Good hops have a lively fragrant smell, and when rubbed between the fingers, leave a disagreeable clamminess not unlike that occasioned by brown sugar.

They are employed to regulate, not promote vinous fermentation; a pure malt-wort would complete its vinous fermentation upon the first experience of warm weather, and pass rapidly into vinegar.

The common country hops are seldom well saved; those from New England should be preferred whenever they can be obtained.

The time of picking, the mode of curing, the care in bagging, and the place of keeping, all have their share in the preservation or destruction of the finer qualities of this vegetable. If the hop be plucked too early, the consequence of immaturity is obvious; if it hang too late, the constant avolation of its fine unctuous parts, wastes its fragrance, destroys its colour, and renders it of less value and efficacy. An application of too much heat in the curing has similar effects, for by evaporating the aqueous parts of the vegetable too hastily, the finer parts of the essential oil rise with them, and are lost, whilst the remainder receives an injury somewhat similar to that of malt by the like injudicious treatment.

The care in bagging and keeping is equally important, on the same principle of excluding, as much as possible, the action of the external air upon the hop, which carries off its more valuable qualities in the same manner as by a too long continuance on the plant. The closer they are pressed down in the bag, the more effectual is their security against this injury, and the best method of keeping them is in a close but dry room, the bags laid upon each other, and the interstices well filled with a dry inodorous matter, such as the first screenings of malt, &c.

Several qualities reside together in the hop, which however, may be separated, and the aromatic flavour, and grateful, mild, astringent bitter, may be extracted alone, without the austere nauseous one, which is produced by long boiling.

For experience has proved, that the aromatic flavour of the hop was extracted by the gentle heat of *infusion* in hot water, that upon a quarter of an hour's boiling, the pleasant bitterness, and the agreeable part of the astringency, came next, and when the boiling was continued about an hour, the nauseous, austere, acrimonious roughness became perceptible, and shewed itself more and more, as the boiling continued, while the aromatic, grateful parts, after boiling an hour, began to be volatilized, and were soon after evaporated and lost, till at length, nothing but a mere bitter, like that of *quassia*, but much more nauseous, was all that was left.

A patent has been granted in England to Henry Tickell, for a mode of preserving the essential oil of hops. A particular description of the machinery may be found in the *Repertory of Arts*. I do not consider it worth transcribing.

[The following publication from a very respectable gentleman in one of the eastern states, contains so much important information respecting the culture of the Hop, that I will be readily excused for its insertion.]

"The intrinsick value and extensive use of the *Humulus lapulus* or *hop*, is so universally known, and all its excellent qualities converted to their full account with so much simplicity and ease, that it excites our surprise to see the cultivation of this inestimable vine so little understood, or so much neglected, especially in those sections of the country where it is most used, commanding the highest price, and in a soil where it would flourish to the highest decree of productiveness. In New England, particularly in the state of Massachusetts, where the hop is cultivated with great success, on an extensive scale, the soil is far inferior to those of the western or middle states; yet in one county (Middlesex) and a few towns of another (Essex) there are undoubtedly more hops raised than in any other five states, or even perhaps the whole union. The vine is cultivated with much more labour and expense, and yet the produce sells at a price less than in those states where it is most used, and might be cultivated on the most advantageous scale, not only as an article of domestick utility, but a staple commodity for exportation.

"The hop is of the reptile species, and is sexified; the female flower being far superior to the male, both in size, weight and quality. Its culture is simple and easy when well understood, and will render a profit, even in Massachusetts, at 12 ½ cents per pound.

"The ground selected for a hop field should be a dry rich meadow or river flat, as far from the stream as may be, but never ascending a step. If the field has been under a growth or crop, it should be well ploughed in the month of October, and the roots placed the distance of four feet apart: four roots, in parts containing one or more joints, placed in each hill at a convenient distance from each other, and the whole covered with earth or manure. In April, or as soon as the frosts have subsided, the mounds are to be carefully opened, as the young shoot is very tender. The vine will then be suffered to grow, the plough having passed at right angles across the field, until the vine shall have acquired a sufficient length and strength to ascend the poles. Poles must then be set, two or three in a hill, according to the appearance of the vines." The best vines are then very carefully wound spirally up the poles, with the course of the sun; no more than three healthy vines to a pole; the residue are suffered to languish; those which ascend being secured by a thread, generally in two or more places, as the poles are of length, say from 12 to 20 feet.

"The hop field is generally ploughed, and hoed from three to four times, first at the time of poleing, and last when the flower bells, which will be in August, either earlier or later, according to the season and climate. It ripens in all September, and which is known by the seed, which changes its orange colour to a brown smoke; it is then gathered with the utmost expedition, as the equinoctial storms destroy the flower when they occur at this state of the ripened top."

Chapter XIX.

To make four Gallons from the Bushel.

(From *McHarry's Distiller.*)

"This is a method of mashing that I must approve of, and recommend to all whiskey distillers to try it; it is easy in process, and is very little more trouble than the common method, and may be done in every way of mashing, as well with corn as rye, or a mixture of each, for eight months in the year; and for the other four, is worth the trouble of following. I do not mean to say that the quantity of four gallons can be had at an average, in every distillery, with every sort of grab, and water, and during every vicissitude of weather, and by every distiller; but this far I will venture to say, that a still house that is kept in good order, with good water, grain well chopped, good malt, hops, and above all, good yeast; together with an apt, careful, and industrious distiller, cannot fail to produce at an average for eight months in the year, three and three quarter gallons from the bushel, at a moderate calculation. I have known it sometimes to produce four and an half gallons to the bushel, for two or three days, and sometimes for as many weeks, when perhaps, the third or fourth day, or week, it would scarcely yield three gallons; a change we must account for, in a change of weather, the water, or the neglect or ignorance of the distiller. For instance, we know that four gallons of whiskey is in a bushel of rye or corn, certain, that this quantity has been made from the bushel; then why not always? Because, is the answer, there is something wrong, sour yeast, or hogsheads, neglect of duty in the distiller, change of grain or change of weather, then of course, it is the duty of the distiller to guard against all these causes as near as he can. The following method, if it does not produce in every distillery the quantity above mentioned, will certainly produce more whiskey from the bushel, than any other mode I have ever known pursued.

"Mash your grain in the method that you find will yield you most whiskey; the day before you intend mashing, have a clean hogshead set in a convenient part of the distillery; when your singling still is run off, take the head off and fill her up with clean water, let her stand half an hour, to let the thick part settle to the bottom, which it will do when settled; dip out with a gallon or pail and fill the clean hogshead half full, let the hogshead stand until it cool a little, so that when you fill it up with cool water, it will be about milk warm; then yeast it off with the yeast* for making four gallons to the bushel, then cover it close, and let it work or ferment until the day following, when you

*No directions are given for making this yeast.

are going to cool off; when the cold water is running into your hogshead of mashed stuff, take the third of this hogshead to every hogshead (the above being calculated for three hogsheads) to be mashed every day, stirring the hogsheads well before you yeast them off."

Remarks by H. H.

The foregoing is published, because the writer says he has practised it with success; I must however, differ with him as to "ease" with which this plan may be adopted. With a patent still it is wholly impracticable, and with the other stills, it certainly is a very troublesome piece of business to empty a still with a gallon measure. The time taken up to do this, added to the time necessary for the still to cool, will be so great, that only two instead of three charges can be run a day.

Somewhat similar to the above, is the following plan. It is however much easier, and I think will be found equally profitable:—Have a small cistern placed over the swill cisterns, so high as to be over the tops of the mashing hogsheads in the house; when the swill becomes cool, pump up with a small pump, to be provided for the purpose, about 150 gallons of the top or thinnest of the swill into the upper cistern, of this put twenty or twenty-five gallons into each hogshead at cooling off; this is called the *returns*.

Chapter XX.

Process by which 800 to 820 pounds of corn, yield 432 to 448 quarts of the best brandy.

Take 800 to 820 pounds of corn good weight, and make it sprout; for this purpose either pure wheat or rye may be employed, or add one-third of malted barly, and mix them before they are ground.

To form the first paste after the corn is ground, the water should not be hotter than 30° Reaumur, or 100° Fahrenheit. If that liquid contain carbonic acid it should be heated to ebullition, and afterwards suffered to cool to the degree specified. When the water contains none of that acid, mix three parts of that liquid, when boiling, with one part of cold water, and let it stand to cool a few moments. Take three parts of this water to two parts of ground corn, according to a measure corresponding to the weight, and mix the water well with the corn, to form an equal paste free from lumps. In this manner the first distillation is begun in autumn; for the succeeding ones, the water remaining from the first distillation is used, after being left to cool to 38°, and then the paste is made with it. This water augments the quantity of brandy produced.

For diluting employ boiling water; pour it slowly upon the paste, continuing to beat it up; add five parts of water to one of paste, which makes the mixture consist of eight parts of that liquid and two parts of corn. Then cover the tub or vat containing the matter, and leave it covered for two hours. At the expiration of that time, uncover it for four hours, for the purpose of reducing the heat, and the cooling may be promoted by stirring the matter every few minutes.

After this add ten parts more of cold water, and beat up the whole with care. This addition will give the preparation the necessary degree of dilution, and will reduce it to a temperature that cannot exceed 17½°, nor be under 16°of Reaumur. Instead of water, as much clear liquor as the operator has at his disposal may be employed. For this purpose draw off, after every first distillation, all the liquor above the pulp, and let it cool in large tubs. This clear liquor considerably augments the quantity of brandy that is obtained.

Then add 60 lbs yeast to 800 lbs of corn. To prevent the sour portion which remains in the vat, and which it is impossible to take out, from turning the whole mixture sour, and to disengage a certain quantity of carbonic acid, add 1½ ounce of good potash, or a pound of good wood ashes sifted, and well beat up the matter.

The vat must then be covered for eighteen hours; after which half uncover it, and when the temperature encreases uncover it entirely after six hours more; if attention were not paid to this particular, a great part of the spirit would be volatilized. If a white froth appear upon the surface it is a sign that the fermentation proceeds as it ought.

Forty-eight hours after the mixture of the paste, the pulp begins to rise; this disposition is promoted by gently stirring the matter, and covering the vat a second time. In seventy-two to ninety hours the whole becomes clear and the fermentation is finished.

The vats must not be larger than to supply one, or at most two distillations. The distilled liquor is rectified every day to half proof, and only to whole proof when there is sufficient to fill the still.

For distilling in summer take a sixth part less of corn to the quantity of water prescribed; and after the addition of the cold water, the degree of temperature should not exceed 8 or 9° of Reaumur.

Chapter XXI.

Process of Distillation in Ireland.

The following account of the process of distillation, is taken from the deposition of James Forbes, esq. of Dublin, for many years concerned in a large distillery. It is extracted from the report of a committee appointed for certain purposes by the House of Commons, and published by their order.

As it gives a complete view of the process of the distillery in Ireland, so very different from ours, I have been induced to insert it.

"The corn is first ground, then mashed with water, and the worts, after being cooled are set for fermentation, to promote which, a quantity of barm is added to them, and they become wash; the wash is then passed through the still, and makes singlings, and they being again passed through the still, produce spirits; the latter part of this running being weak, is called *feints*; when the singlings are put into the still a small quantity of soap is added, to prevent the still from running foul; a desert-spoon full of vitriol, well mixed with oil, is put into a puncheon of spirits, to make them shew a bead when reduced with water; this is only done with spirits intended for home consumption, and no vitriol is used in any other part of the process. In this distillery the former practice was to use about one-fourth part of malt, and the remainder a mixture of ground oats, and barley and oat-meal; latterly the custom has been to use only as much as would prevent the kieve (neach vat) from setting. He had found that malt alone produced a greater quantity of spirits than the mixture of malt and raw corn, of the same quality with that of which the malt had been made. He generally put from 50 to 54 gallons of water to every barrel of corn, of twelve stone (14 lbs to the stone). Each brewing was divided into three mashings, nearly equal; the produce of the two first was put into the fermenting backs, and the produce of the last, which was small worts, was put into the copper for the purpose of being heated, and used as water to the next day's brewing, when as much water was added, as would make, with the small worts of that brewing, 54 gallons to each barrel of corn. The kieves were so tabulated that he always knew the quantity of worts that would come off at each mashing. Their strength he ascertained by Saun-der's saccharometer, and at the above proportions he obtained from a mixture of the two first worts, an increase of gravity from 20 to 22 pounds per barrel, of 36 gallons, above waterproof, at a temperature of about 88. The small worts gained at the same temperature, about six pounds. The grain, after the last worts were taken off, retained nearly the same bulk as when put into the kieve; the whole of the grain was put in at the first mashing, he never knew any grain to be added to the second mashing. The worts of the first and second mashings were run through the mash kieve into the underback, in which state they were usually found to correspond with the computation made in the mash kieve and underback, in the latter of which a correct guage might be taken of them. He usually commenced brewing about six o'clock in the morning; the first worts were run off into the underbacks, and required from an hour to an hour and a half to be forced up into the cooler. The second worts came off at the end of two hours from the discharge of the first, and required about the same time to pass into the coolers. The small worts were generally run off late at night, and being then, or early on the following morning, put into the copper to be used for the next brewing, were seldom shewn on the coolers. He thinks that any decrease of the worts by evaporation, whilst on the coolers, must have been very inconsiderable, and that a correct guage of the worts might be taken as well in the coolers as in the underbacks. The quantity of wash in the backs was found to be nearly correspondent with that of the strong waters which had been on the kieve and in the cooler. The fermentation of the worts was produced by means of yeast, and was in general so contrived as to be apparently kept up for the full time allowed by law (six days): he has, however, usually had his worts ready for the still in twenty-four hours from the time in which it was set. Backs are renewed in two ways, either by additions made to them from other backs in the distillery, each supplying a certain portion of wash to the back which is next before it in the order of fermentation, while the newest and least fermented wash is replenished by worts, or when the fermentation is down, by an entire substitution of worts. He has ordinarily in course of work, charged a 500 gallon still with wash, and run it off in from 20 to 23 minutes. He has seen a 1,000 gallon still charged and worked off in 28 or 30 minutes. He understands that it is now the practice of some distillers to heat the wash nearly to the state of boiling, before the still is charged with it, by which means he believes the process to be accelerated by three or four minutes. He has seen a 1,000 gallon still charged with singlings, and worked off in from 40 to 50 minutes, and thinks a 500 gallon still requires nearly the same time. Feints from pot-ale (the name given to completely fermented wort) usually are run off in from six to seven minutes, making allowance for every delay; about six charges of spirit may be run off from a still of 500 gallons content, each charge estimated at 150 gallons. The feints were always put back into the pot-ale receiver; 20 gallons of feints is the usual quantity run from a 500 gallon still charged with singlings; he thinks there is more spirits extracted from feints than from pot-ale. There was no delay between one charge of pot-ale and another, or between one of singlings and another, the still could be cleansed in less than a minute; it very rarely occurred that the ordinary accidents which happened to the still delayed the work to any considerable degree. The still is never charged with wash beyond about seven-eighths of the still, nor with singlings beyond about four-fifths, exclusive of the head. The estimated produce (according to which the duty maybe charged) is one gallon of singlings from three gallons of wash, and one gallon spirits from three gallons of singlings, but it is frequently somewhat more. Previous to the regulation [the excise] which took place in June 1806, from a still of 540 gallons, which is charged with 2,075 gallons of spirits weekly, he has frequently drawn 5,300 gallons in one week, and thinks 5,000 to be a fair average.

Chapter XXII.

Of Lutes.

It would seem almost unnecessary to make any observations on the necessity of a proper kind of luting, or paste to prevent the loss of liquor, by flying off in the form of steam. This however is a part of the operation which requires close attention.

It is too common with distillers, in many parts of the country to make use of any kind of clay or loam that may be found about the distillery, giving as a reason that it is too expensive to use rye meal; thus they frequently lose as much in one day as would supply a proper lute for twelve months. For, so soon as the clay becomes hardened, it cracks in a hundred places, and the steam pours out at every opening.

For all the operations of a grain distillery, both singling and doubling, paste made of rye meal and water, well worked in the hand, will be found not only to be the best, but the most economical that can be used; but for the nicer operations of a cordial distillery, or in the making of spirits of wine, where the subject disengaged is of a very volatile nature, a stronger luting than the above will be found necessary. The cheapest and most simple of this kind, may be made by a mixture of quicklime with the whites of eggs, worked to a proper consistency: this must be used as soon as made, at it dries almost immediately.

Chapter XXIII.

Observations on the advantage of preparing Whiskey for market, of a proper strength; and of the mode of inspecting Whiskey.

Of the numerous plans adopted by distillers to increase their produce, there is none so improper in itself, so deceptive to the purchaser, or eventually injurious to the distiller himself, in any situation, as that of putting a false bead upon gin and whiskey.

To correct this evil, to place the ignorant purchaser, upon a footing with a more skilful one, and to do justice to the distiller who sends his spirit to market of a proper strength, the corporation of the city of Baltimore in the year 1807, passed an ordinance subjecting to inspection, all gin and whiskey which might be offered for sale in that city. Similar ordinances are now in force in New-York and Philadelphia.

Notwithstanding this regulation, many distillers continued to send their whiskey to market considerably below proof, under a belief that it was the most advantageous plan. To correct this mistaken notion a calculation is subjoined of the expenses and net proceeds upon the quantity made in one year, if sent to market of a proper strength; also a calculation of the same of reduced ten degrees below proof.

Suppose a distiller to be situated 60 or 70 miles from market, and makes in the year, 8,000 gal, worth at market say 50 cents per gallon.		. $4,000.00
Supposing barrels to hold 30 gal each, it will require for the above 266 barrels.	$266.00	
Hauling 8,000 gal at 6 cents	. 480.00	
Gauging and inspecting 266 barrels at 12 ½ cents .	. 33.25	
		[779.25]
Net proceeds will equal		. $3,220.75
Now supposing the above to be reduced 10 per cent, it will amount to 8,800 gal, which at 40 cents per gallon will amount to		. $3,520.00

At 30 gal per barrel it will require 293 barrels . . .	.293.00	
Hauling 8,800 gal at 6 cents	.528.00	
Gauging and inspecting 293 barrels at 12½ cents . .	. 36.62½	
		[857.62½]
Net proceeds equal		. $2,662.37½
To be deducted from net proceeds of proof whiskey as above . .		3,220.75
Leaves the balance in favour of proof whiskey		$558.37½

Now if a calculation be made upon the above principles, supposing the price of proof whiskey to be 40 cents per gallon, which indeed will be generally found more correct, the advantage in favour of proof whiskey will be still greater.

Is not this worth attention, particularly as the manufacturer of strong whiskey will find a preference always given to him?

We will now suppose the distiller to make his whiskey ten degrees above proof.

He will then have 7,200 gal, at 60 cents		. $4,320.00
240 barrels will hold it	$240.00	
Hauling 7,200 gal, at 6 cents	.432.00	
Gauging and inspecting	. 30.00	
		[702.00]
Net proceeds.		. $3,618.00
But the net proceeds of the proof whiskey was		. [3,220.75]
The balance then in favour of this		397.25
Added to		558.37½
Shews the difference between making whiskey 10 degrees below or 10 degrees above proof, in favour of the latter		$955.62½

Extend this calculation to spirits of wine, and the proportion will still be found favourable to the stronger spirit. But if it be made fourth proof, and proper attention be paid to the quality, it may be readily sold for one dollar per gallon.

The above calculations deserve the serious attention of every distiller. From them it is evident that by reducing the proof, although the bulk or quantity is increased, the expenses are also increased, but the price is decreased; the latter will be found greater in proportion to the increase of the quantity.

But as the proof is raised, the proportion increases in his favour.*

*The act of the legislature of New-York requires that all spirituous liquors offered for sale in the city of New-York, be subject to inspection. Southworth's hydrometer is the standard, by which first proof is fixed at 15 degrees below proof on Dicas's hydrometer.

Contracts for gin or whiskey are understood to be for hydrometer proof, and the inspector marks accurately the degrees under or over proof, for which an allowance is always made.

The act of the Pennsylvania legislature requires that all spirituous liquors shall, prior to exportation from the port of Philadelphia, be gauged, inspected, &c. Liquor 10 degrees below proof on Dicas's hydrometer, to be considered as 1st proof—5 degrees as 2d proof, and 5 above as 4th proof. Dicas's hydrometer is the standard. In all cases where the spirit is not proof, the inspector's fee shall be paid by the person offering the same for sale; in other instances it must be paid by the purchaser.

The mode of inspection in Baltimore is not generally known; a few words therefore are necessary to put distillers on their guard on this subject.

When gin or whiskey is below proof, every degree is carefully marked, and a deduction of one cent per gallon for each degree, will be made by the purchaser. But if it be a few degrees above proof, no notice is taken of the number of degrees, except they be sufficient to entitle it to be called 2d proof, which may be 9 or even 12 degrees above proof; and 4th proof spirit, which is 25 degrees and more indeed above proof, will be marked about 15 degrees above proof, at the will of the inspector.

Chapter XXIV.

Of Geneva or Gin.

Gin is formed by the distillation of whiskey, or any other spirit, with a portion of juniper berries, or oil of juniper.

The name of Geneva is derived from *Genevre* the French name of the juniper berry. It was formerly a custom to mix a variety of article with malt spirits, in order to take off the disagreeable flavour. Among other things used for this purpose, some tried the juniper berries, and finding that they gave not only an agreeable flavour, but very valuable virtues to the spirit, the custom became general, and the liquor sold under this name. The method of adding the berries was, to the malt in grinding, a proper proportion was allowed, and the whole was reduced to meal together, and worked in the common way. The spirit thus obtained, was flavoured *ab origine* with the berries, and exceeded all that could be made by any ether method.

It was soon however discovered, that there was a great similarity in the flavour of oil of juniper and that of turpentine, though a very material difference in the price; the juniper was accordingly, entirely omitted, and spirits of turpentine substituted; such is the common practice in England, at present.

I have never heard of the method of grinding the berries, being used in this country, though there is no doubt, that by mixing the oil or berry, with the wash previous to, or during fermentation, a more complete union will take place.

The use of spirits of turpentine however, has unfortunately become too common, and is one great cause of the badness of American gin, and consequent prejudice against it. But as this article is frequently mixed with, and sold for, juniper oil, the distiller is deceived, and at a loss to account for the bad quality of his gin.

The improvements in gin making, have been very considerable within a few years past, and some of our distillers seem to be actuated by a laudable determination to equal the Holland gin, justly esteemed superior to that of any other part of the world. This is certainly a great desideratum, when we consider that Holland gin is selling for from two dollars to two dollars and fifty cents per gallon, while our American gin is frequently sold at fifty cents, or if of very superior quality, ninety cents.

Whence this great difference? and why cannot we make gin equal to Holland? The superiority of their gin is generally attributed to some *secret*, known only to themselves, and which has never got without the walls of their distilleries.

That we may make gin equal to theirs, without the aid of this wonderful secret, I believe, and for these reasons.

It is a well known fact, that they frequently have taken grain from this country to make gin; it is also known, that they use large quantities of buckwheat which is considered inferior to rye or corn; it is well known that there is a great difference in gin from the different distilleries in Holland, and it is also well known that distillers from Holland who come to this country, cannot make better gin than we do ourselves. Many who have experienced the fact, state, that the new gin in Holland, is not better than that made in some distilleries in this country; and it has also been ascertained that a voyage to sea has so improved American gin that it passed for Holland. What then but age and the sea voyage creates the difference? Or have they a different method of incorporating the juniper with the spirit? And are they not more attentive to *cleanliness** in their distilleries than we ? With respect to the great *secret*, if this is a fact, would not some of the distillers who emigrate to this country bring it with them? Or, would not some American ere this have obtained it either by purchase or by bringing over some person acquainted with it? Whence the difference in quality of Holland gin, from different distilleries, if they *all* have the same *secret*? Let those who argue in favour of the secret, and who say we cannot equal Holland gin, take into consideration, that fifteen or twenty years ago scarcely any attempts were made in this country to manufacture gin; that the Hollanders have pursued it as a regular business for 400 years, and that in the single town of Schiedam in the year 1775, there were 120 distilleries; in 1792, there were 220; and in 1798, there were 260; and in the whole province of Holland, 400; each of which made annually 4,992 ankers or pipes of gin.

These distilleries were probably the sole dependence of their owners for a living; and to them their whole attention was devoted. But in this country, even to this time, how few distillers are there who depend solely upon their distilleries? or who pay any attention to the *quality* of their spirit?

Let it also be recollected that gin is a product of art, and does not depend, either upon soil or climate; and that if the same materials be employed, and the same process be observed, the result must be the same, whether in Russia, Holland or America.

Their materials we know to be the same, and their process corresponds nearly with ours. But there are many little circumstances which affect the quality of gin, to which we pay no attention.

Let us not, then, because some few men have failed to imitate Holland gin, by pursuing what they *supposed* to be the Holland plan, give it up as impracticable; but rather, reflecting, that as it requires time to bring any art to perfection, we *must* ultimately succeed by pursuing it steadily with that determination.

What would we say of a bungling mechanician who, because he failed in finishing off an article equal to an imported one, pronounced his pattern *inimitable*?

I have indulged myself in these observations from a belief in their correctness, and in the hope that it may stimulate our distillers to pay more attention to a matter not absolutely hopeless; but to return to the subject more immediately requiring our attention.

*The cleanliness of the Hollanders is proverbial.

There is certainly an advantage in rectifying whiskey previous to its distillation with juniper, this however is too tedious and complicated an operation for the grain distiller. I have generally succeeded very well in making gin, and the plan I adopted, was to throw into each charge of singlings a sufficient quantity of juniper berries, without any other addition whatever, being satisfied that most of the other ingredients generally used are rather injurious than beneficial; the quantity of berries can only be ascertained by experiment, as there is considerable difference in the quality; generally however, 20 and 30 pounds to 110 gallons spirits, will be right.

Since the above was written I have read in the *Archives of useful Knowledge*, a paper on the subject of the gin distillery in Holland, by a Mr. Crookens. Some of his assertions appear to me erroneous, and it is difficult to reconcile others. He notices however, the difference in the quality of gin made at different distilleries, and lays much stress upon the kind of water which is used, and the manner of making and using yeast, which he says is kept a *secret*. He says they prefer rye grown upon a dry sandy soil, and mostly use the Prussian rye, which from the circumstance of being kiln dried, is called dried rye, and at least one fourth malt. This accounts for the difference between Holland and American gin in some measure, for it has always been observed that there was a *rawness* in American gin not perceptible in Holland gin, supposed to be from the use of com; and to divest it of this flavour has been always considered necessary to imitate Holland gin.

From experiments however which have been made with rectified whiskey, it is evident that something more is necessary than merely to divest our spirit of this raw flavour. It is also requisite to give some other flavour with the juniper. This must probably be done during the fermentation, and by using kiln dried grain. *Care, attention and cleanliness, age and sea voyage,* all assist.

It has been asserted by Dutch distillers who come to this country, that the Holland distillers use *dry yeast*, which is sent down the Rhine from the German breweries.

The scarcity and high price of berries has obliged the distiller to resort to the use of the oil of juniper, which has been reprobated for the reasons above stated: unfortunately the distiller has no method of detecting the imposition, and is equally deceived with the consumer; when however genuine oil can be obtained, it will be found equal to the berry.

Shaw in his essay on distillation, says, "the best method of introducing the oil, so as to avoid all inconvenience, is to reduce it first to an oleosaccharum by grinding it in a mortar with a due quantity of fine sugar in powder. The oil thus added, with its particles disunited and in form of powder, will readily mix with the liquor (or wash) and immediately ferment with it."

The method which I have adopted in using oil is this:

Take two gills of juniper oil, pour it on four or five handfulls of rye meal, and stir it until it has the appearance of brown sugar, then pour on boiling water enough to make it of the consistence of paste, add rye meal to make it into stiff dry balls, each containing sufficient for one charge of singlings, the whole being equal to eighty pounds of berries, provided the oil is good; a few trials will shew the proper quantity.

When the oil of juniper can be obtained perfectly pure and unadulterated, it may be mixed with alcohol. After standing a few days until the oil is completely dissolved and united with alcohol, it may be mixed with proof spirit. I have known this to be done, and the gin thus made, not to turn blue upon being mixed with water; it however cannot be relied on owing to the impurity of the oil, except obtained direct from the importer.

This information well known to chemists, is *sold* by many pretended Holland distillers, as a great *secret*.

Chapter XXV.

Of the advantages of feeding Swine or Cattle, and the proper management thereof.

The wash or swill after distillation, affords good food for hogs, or cattle, and if properly managed, this branch of business, will be found to form a considerable item in the profits of a distillery. Yet it is not only neglected by many distillers, but even said to be unprofitable. Setting out with the absurd idea that hogs will not thrive, but in dirt, and that they are naturally fond of living in the mire, a number of hogs are crouded together, in a small pen, without a floor, it is soon rooted up in every part, and upon the first rain becomes a perfect quagmire. The trough out of which they eat, is generally half full of mud, and the swill is let into it, in a half boiling state. Yet even this unpalatable mixture, the natural voraciousness of the animal, induces it to swallow. The inevitable consequence of this treatment is, that the animal becomes diseased and falls away; yet will frequently continue to eat as heartily as ever, which induces the owner, without once reflecting that there is any thing wrong or wanting on his part, to conclude, that swine are unprofitable stock for a distillery. In cases where rye is used alone in the distillery, the swill is not so good as from a mixture with corn; hence it will be observed that in all rye countries the feeding of hogs is not considered so profitable as in places where much corn is raised. It is questionable whether hogs wallow in the mire from love of filth, from a desire of taking a bath, or of being cooled; for it is observed that in warm weather they will root up the fresh earth when they wish to lie down, or will go into a creek and sit for a considerable time completely immersed in the water, except the head. The writer of this, had a pen so situated that he could bring water six or eight feet over the top, where it was conducted in a small trough. It was suffered to run for a few hours every day in very warm weather. As soon as it commenced, the hogs would arrange themselves around, and one at a time get under the spout; after remaining a short time and being well washed, they retired to a clean part of the pen to lie down, and never shewed a disposition to wallow in filth while the clean bath was daily given. Hogs will at all times, when practicable, keep a part of their pen clean and dry for sleeping on, and carefully avoid the dirty part. Hence it may be concluded that hogs require a warm, clean, and dry pen.

The situation for the pens, therefore, should be considered in the first erection of the distillery. That they may be, as far as practicable, protected from the northern and eastern storms, exposed to the sun, and that there be sufficient fall for the swill to run from the cistern into the troughs, out of which the hogs feed.

The pens should be built with substantial plank floors (two inch oak, if possible, as hogs will sometimes eat up pine plank), with an elevation of about four inches to every ten feet in width towards the back part of the pen, which should be covered in completely for a protection against the weather. The troughs must be so contrived as to prevent hogs from getting into it. This is most effectually done by means of rounds, let in with an inch augur, about two inches from the top of the trough. These are less liable to be destroyed than slats nailed across the top, and if done as the trough is first put together, will render it much stronger than it would otherwise be. The rounds should not be more than eight or nine inches apart. The trough about nine inches deep, and wider at top than the bottom. The pens are to be made of proper size, for about fifteen hogs each, allowing about ten square feet for each hog, a sliding door to be made for each pen; communicating doors between the pens will also be found very convenient.

The pens being prepared, care should be taken in selecting the stock of hogs proper for the establishment. In doing this, none must be admitted of less than 60 pounds weight; and sickly or diseased hogs, by no means allowed. They should be classed in sizes as near as may be conveniently done, and two or three more put into each pen, than the proper complement, for it is frequently necessary to take out two or three during the season, who get hurt or become diseased.

The swill with which they are now fed should be allowed to get quite cold before it is given to them, which is easily managed by having two separate cisterns to be used alternately. Pieces of soft brick bats should occasionally be thrown to them; and salt is said to be sometimes necessary, but I always found it to injure some of my hogs, when they eat it. The pens should be kept clean, as it is fully evident, from the preceding remarks, that hogs do not delight in dirt, but in cleanliness, without which, at least they certainly will not thrive.

Care should be taken not to have too large a stock; it is better always to have a little swill to spare. The usual calculation is, that one bushel of grain will feed ten hogs. This I think rather exceeds the mark; it can only however be ascertained by experience.

The preceding directions are particularly applicable to the management of hogs that are fattening at a distillery unconnected with a farm. When however a farm constitutes part of the establishment, it becomes a question whether hogs do not thrive faster for occasionally running out, and getting grass.

There is a considerable variety of opinions on this point, but, so far as my own experience goes, and from the best information I can obtain from others, I believe that if proper attention is paid to hogs they will fatten sooner, and upon less food when closely confined than when suffered to run at large.

It is for the purpose of raising pigs, that a farm is particularly valuable, as an appendage to a distillery.

No pains should be spared in getting a good breed of hogs, where it is intended to raise them for the distillery. The most profitable kind are such as grow large and fatten quick; two properties seldom united, but best obtained probably by an union of the large English breed, with the *no bone*, or with the China hogs. This is the best kind I have met with; there are however, in many parts of the country very fine hogs which probably are as profitable as these I mention. A number of sows are to be kept for breeders, and to be well fed at all times; the pigs at three or four weeks old, should be fed with meal and swill, and allowed full liberty to go to the swill when they please; if they can also have the range of a clover field, they will be fit for the pens at two months old. The notion that swill injures them, is erroneous. It is the dirt and confinement, when a number are put into a small dirty pen. Care should be taken, that they are not stunted in the growth, if this happens, it will be a long time before they get over it. Connected with a distillery, there is probably no better way of employing a part of a farm, than in raising pigs. The boar pigs should be castrated quite young, as soon indeed as they can be handled; and the sow pigs spayed at about four months old. They will thrive much faster after this operation, than if it had not been performed.

A sow goes nearly four months with young, and brings forth from 6 to 13 or even 15 at a time. They generally breed twice a year, sometimes, three times, and the following remarkable increase is thought worth notice here: a sow, in England, four years old, has farrowed 229 pigs, which is an average of 57 per year, and except the first time, always brought up 13. Observations upon the importance of this subject might be extended to a considerable length; anything further however is deemed unnecessary, but merely to remark, that with proper care, a hog will gain more than one pound a day. This assertion is doubted by many and even laughed at by others; to satisfy all reasonable people, I take the liberty of subjoining the following account of two prize pigs, shewn before the Smithfield Club, in England, at their annual meeting in December, 1810. Such increase as this however cannot be always expected, but we should endeavour by close attention to approximate it as much as possible. From the best observations I have been able to collect, the increase of each hog at a distillery, taking the general average of some years, is rather more than six-tenths of a pound, a day.

Prize Pigs shewn before the Smithfield Club, in England, in 1810 Annual Meeting—December.

PRIZE PIGS.	Pork & Head.	Loose Fat.	Feet.	Blood.	Entrails.	Wt. alive.
Mr. I. Roads, 60 weeks old, spotted Berkeshire pig, fed on skimmed milk, and four bushels barley meal.	502	14	3	8½	25½	553
Mr. I. Roads, 40 weeks old, spotted Berkeshire pig, fed on skimmed milk, and four bushels barley meal.	366	14	3	5	25½	411½

In 1802 there was a hog in England, in an unfattened state, allowed by competent judges to weigh one hundred stone. It measured in height, between 3 and 4 feet; in girt 8 feet, and in length from the point of the snout to the extreme of the tail, 10 feet. It was sold to the owner for 35 pounds.

Three hogs were sold in Wilmington, in January, 1813, which weighed together, 2,301 pounds.

Mr. Eli Cooley, of Deerfield, (Mass.) fatted eight hogs, which were killed in February, aged 18 months, which weighed as follows: one 590 pounds, one 520, one 502, one 500, one 455, one 452, one 428, one 405; six of which were sold in Boston for $14.62½ per hundred. The Massachusetts Agricultural Society granted him the premium, 50 dollars.

Swill is of great service to milch cows, and cattle may be advantageously fed with it. It is alleged, however, that without hay they will not thrive, as they must have a cud to chew. From my own experience I can say nothing as to this fact, but I have heard of several instances of cattle being put into the pens with pigs, and very soon becoming fat, fine beef, without any thing to eat but swill. It is also stated that cattle will not thrive when confined. The following statement, which may be relied on as correct, will fully satisfy any one of the incorrectness of this opinion. It will also go to shew the incorrectness of the prevalent opinion, that cows cannot bear confinement. It will serve to shew the vast advantage which may be derived from a dairy connected with a distillery, in the neighbourhood of a large city, where a large revenue may be daily received from a number of cows, which are at the same time increasing, in size and fitness, for the market,

Many of the dairys on Long Island, near New-York, are supported from one distillery; very little other food is used. If some of our capitalists would undertake a similar establishment near Philadelphia, it would be to them a source of considerable profit, and a benefit to the public at large.

"In Glasgow one of the greatest curiosities shown to strangers, and one of the greatest curiosities in Britain, is a cow-house, set up on his own plan by a Mr. Buchanan, an old but a very skilful and successful master-weaver. In this cow-house are kept constantly about 300 cows, in the neatest, most clean and healthy order. The house (one room) is a square building, the roof supported in the centre by iron pillars. The floor is boarded, washed clean, and sanded. Small long stages, about a foot above the floor are erected, each containing perhaps twenty cows. These stages are just as wide as the cow is long, and behind the cow is a trough to carry away what falls from it. They are kept two and two together, and fed regularly with grass of some kind, and watered; women attend upon them, and groom them as men do horses; but during the nine months they are in milk they never change their situation. They live upon about six square feet each, yet their skins are always sleek and silky, they are fat and beautiful. The moment they become dry they are sold to the butcher, for whom they are highly fit. The owner has a man or two travelling about the country, purchasing new ones coming into milk; the owner also keeps a farm, which the cows' manure enables him to dress well. In this way the business goes on like clock-work, it being but secondary to his weaving trade, and has gone on for eight years; no bustle, no confusion. He sells his grass milk for half the price the Londoners sell their nauseous mixture, though land is dearer around Glasgow than around London."

Chapter XXVI.

Miscellaneous Observations.

Every man almost who enters into business, of whatsoever kind, has a certain degree of confidence in his own knowledge, or in his abilities to acquire information, that soon induces him to adopt a particular routine, from which it is difficult to persuade him to depart. Others however are convinced that none are too wise to be informed, and that valuable hints may sometimes be obtained from the most ignorant. To such, a few observations, the result of several years practice, are submitted.

The owner of a distillery, who pays proper attention to it, will find a variety of occupations, constantly demanding his superintending care; but unfortunately, distillers are too apt to attach themselves to some particular branch of the business to the neglect of the others. Thus the whole delight of one is to see fine fat hogs; of another, that the fermentation is good; of a third, that there should always be sweet yeast; of a fourth, the manner of running the stills; while a fifth is constantly riding through the country in search of wood and grain (which has the effect of preventing the sellers of these articles from coming to him). The impropriety of this kind of partial management, must be obvious upon very slight reflection. A detail of all the minutiæ, requiring attention would be tedious and unnecessary; they will soon become obvious to a man of observation.

A due regard to his own interest will convince the owner of a distillery of the necessity of having a regular supply of grain, wood, and every thing necessary; no excuse will then be afforded to the hands for the neglect or postponement of their work.

Care should be taken in hiring hands, that they are careful, sober and honest; each should have his particular station; or they should be put under the direction of one, who should have a general superintendance and be accountable for the internal management of the distillery.

A regular discipline should be established, and strictly adhered to. The mashing should commence at a particular hour every day; by which much confusion and loss will be avoided, and the owner may always make his arrangements to be present at this operation, without interfering with other parts of his duty. The doubling still should also be charged at a particular time, and always run in the day time; so soon as the spirit is off, it should be reduced and made ready for market, and the still be washed clean and filled with water, which should remain until the time of charging again the next day.

The advantages of a regular system cannot be too strongly inculcated.

It is a common observation, that any *fool* may be taught to run a still; this *may* be true; the writer of this however, enters his caveat against trusting such; the man who attends a still should be particularly careful, and possessed of sufficient presence of mind to attend to his duty in case of accidents; which may happen in the best regulated distilleries; he should have sufficient sense to perceive at once when any thing is wrong, and act accordingly. The noise of the chains in a patent still are very perceptible, and some little exertion is necessary in stirring; the writer of this however trusted his stills to a *fool* who had been two or three years in a distillery; the spindle broke, but he continued *charging, turning and running* the still for eighteen hours in this situation, until a hole was burnt through the bottom. The smell was so strong as to be perceptible the next day at a distance of one-hundred yards.

Particular attention should be paid to the hogs, that they are kept clean, and regularly fed with *cool* wash.

In some instances we find men trusting their whole establishment to the management of hirelings, by whom they are flattered with the idea of getting three gallons to the bushel, when upon winding up at the end of the year, they find a yield of nine or ten quarts; this is a painful discovery, especially upon hearing others boast of an annual average yield of three and a half to four gallons from the bushel; will not this stimulate to more particular and close attention? Let it also be added that some European chemists have advanced the opinion, and even given directions for obtaining six gallons, or even more from the bushel; however, although this idea may be ridiculed, we certainly do not know the precise quantity of spirit contained in a bushel, or the method of extracting it; nor can it be known but by actual experiment; neither can we set bounds to our expectation, when we consider that one gallon of molasses has been known to yield more than one gallon of spirit. Here then is a wide field for the industry and talents of the distiller, and he who succeeds better than his neighbours, even in a slight dergree, in his experiments, will find himself amply remunerated.

Another thing requiring attention is the flavour of the whiskey. Almost every distilled spirit partakes of the flavour of the subject from which it is distilled; but in fermented liquor, this may be destroyed by any more powerful agent which may happen to be present. Now the flavour of rye, corn, or malt, separately considered, are not disagreeable, and when mixed together in the mash tub, their combined flavour is very pleasant; whence then the nauseous smell of common rye whiskey? Some attribute this to the corn, even in cases where no corn is used! I think however it will rather be found in one or more of the following causes, to wit: dirty mashing hogsheads, bad yeast, dirty stills and worms, and reducing with feints. That the flavour above alluded to is not the natural flavour of grain spirits, is abundantly proved by the superior taste of specimens of well distilled whiskey, which we occasionally meet with. Let those who doubt, or wish to be satisfied, try the following experiment, at a season of the year when they may calculate upon good regular fermentation.

Prepare a sufficient quantity of perfectly sound grain, let the hogsheads be completely scalded, and if necessary, scraped; and just before using, well burnt or fumigated with oats straw; every thing used in the operation, to be perfectly clean; the yeast to be sweet and well hopped; then proceed to mashing in the usual way, taking care to keep every thing acid from the hogsheads. The mash, and (during the fermentation) the wash, will be strongly impregnated *by the hogshead*, with the flavour of the fumigation; and the spirit, if distilled in a clean still, &c. will be perfectly free from the nauseous flavour so much complained of, and partake slightly of that of the cask, provided no feints are mixed with it.

The collection of verdigris in the worm, should be carefully guarded against, to do this, the worm should occasionally be filled with hot swill, and remain so for some time, after which it should be washed with cold water; this should always be done, when the still is stopped for any time exceeding twelve hours.

If the globe of a patent still be not frequently cleansed there will be found adhering to it an oily substance, which may be properly termed the concrete oil of rye and corn.

This certainly dissolves the copper, as may be seen by rubbing it with the finger, and is one great cause of the impurity of gin and whiskey. If suffered to collect for about three weeks, it will then be found coming through the worm with the singlings, and though not at all times perceptible, will be very evident if the singlings be run through a flannel strainer, a precaution very necessary to obtain good spirit. Even where the utmost attention is paid to cleanliness, a fine flannel strainer will separate and retain a portion of the essential oil. *Clean* tow will answer the same purpose.

It is not uncommon to hear distillers talk of *letting their casks sour*, instead of ferment; and indeed not to distil them, until they are sour, which however is the case very soon after the fermentation subsides; and is a sure evidence that something is wrong in the process, vessels, or materials. Such casks never yield much spirit. One of the causes is bad mashing. A sufficient quantity of saccharine matter, not being extracted in proportion to the water used, the wort is not sufficiently strong to keep up a regular vinous fermentation throughout, and the acetous takes place too soon. But where a regular and perfect vinous fermentation takes place, little or no acidity is perceptible, nor will it appear for several days, except in very warm weather.

The thermometer and hydrometer should always be at hand, and frequently resorted to.

No certainty can be attained without a notation of circumstances that alter the product occasionally, and this cannot be correctly made without reference to these instruments.

How frequently is a distiller prevented from making an experiment, or a slight alteration in his usual process, because he has no mode of ascertaining the product of a particular cask. And how often does he wish to know the produce of some one cask, and can only do it by guessing? It is true, he may measure precisely the quantity of singlings, but of their strength, he cannot judge. He cannot compare one with another, for a cask which yields 20 gallons of singlings will yield probably double as much spirit as is produced by only 13 or 14 gallons from another cask. This will be accurately shewn by the hydrometer. The produce of every doubling may also be ascertained before it is run.

The satisfaction arising from this is great, and still greater must it be, to have a correct account of his experiments or daily work, to which he may at any time refer.

The importance of these things will be made more evident from the following circumstance, which took place in my distillery.

In the spring of the year 1808, I had occasion to change my distiller, in consequence of which, a thorough cleansing took place of every thing in the house, the casks were burnt with straw, and fresh yeast made. Having other business to attend to, occasionally, I did not pay particular attention to the mashing, but I observed the fermentation to be very good. In four days were mashed, according to the directions in the 12th chapter of this work, 20 hogsheads, containing 20 bushels of corn, 10 bushels of rye, and between 3 and 4 of malt. The produce was 115 gallons of strong gin, above proof, which sold in Baltimore for one dollar a gallon.

Had a thermometer been used, and a due degree of attention paid to every part of the process, I would probably have discovered to what circumstance this increase was owing. To common observation, there was no perceptible departure from the usual mode of mashing, daily pursued. But I have not been able since to obtain a similar produce. Accurately estimated, it would have probably been equal to four gallons to the bushel, of such proof as is usually sold in Philadelphia.

In making gin or whiskey ready for market, the hydrometer is useful. Indeed, to the distiller whose whiskey or gin may be subjected to inspection, it is indispensable, and he will find that the advantage derived from its use will soon repay the expense and trouble.

The casks in which the gin or whiskey is put, also deserve attention, for it will sometimes be found that when, after great care and attention a spirit is made, which is of fine flavour, after standing a few weeks it is quite altered, and becomes disagreeable; this must be from the cask, and shews the necessity of preventing a similar accident in future.

It is said that Holland gin never tastes of the cask, and, that *something* is put into the cask to prevent it from imparting any colour to the gin; and this is also a *secret*! Let us examine the fact. It is well known that oak wood contains a colouring matter and what is called an acrid principle; the former of which gives colour, the latter taste, but both of these can be extracted, or evaporated by seasoning; consequently if staves are sufficiently seasoned, before they are made up, the cask cannot give either colour or taste to any thing which may be put into it. But it is too frequently the case in our country, that half seasoned wood is used, and the consequence must inevitably be, that when the casks are filled with spirit, it gradually extracts the colouring matter and acrid principle, and the latter injures the flavour of the spirit. Hence the necessity of making new casks of perfectly seasoned staves, as they do in Holland. As the cask always partakes of the flavour of the spirit put into it, this must be attended to in using casks of second hand; therefore, gin should not be put into wine or brandy pipes, or rum puncheons, as well on account of flavour, as of the colour which it will certainly receive. But as colour is not objected to in whiskey and its flavour may be thereby inproved, it may be put into the pipes, but not puncheons, as they render whiskey disagreeable. Cider, beer, or vinegar casks, always injure spirit; these things are well known to dealers in spirit, and the high price frequently given for gin pipes on account of the flavour, is also well known.

Of Accidents which may occur in distillation.

It may not be improper here to observe that accidents sometimes occur, the least consequence of which is the failure of the operation and the loss of the ingredients, or charge in the still.

In a grain distillery and with a patent still, the stirrers sometimes get out of order, which may be easily discovered by the difference in turning, or cessation of the noise made by the chains; if this is not noticed in time, the wash adheres to the bottom of the still and burns; a strong disagreeable smell now becomes very evident and the bottom of the still where burnt, appears perfectly white outside. Whenever this is discovered, the fire should be put out, the still discharged and filled with cold water; when sufficiently cool the still must be examined, and a black crust will be found sticking to the still; this must be carefully scraped off, and the still rubbed until perfectly bright; if this precaution is not used, a hole will be made in the still in a few hours.

If the fire be too violent, particularly when the still first comes round, the wash rises into the head, runs foul and sometimes chokes the worm. This is to be avoided by careful attention to the fire, at all times and by damping when too great. If this is not done in time the head will burst, as it cannot fly off; the consequence of which would be terrible to any one unfortunately within reach of the wash.

But it is from the doubling still that the greatest danger is to be apprehended, as the spirits may take fire, and if a remedy be not applied in time, burn down the distillery.

These accidents are only to be prevented by constant care and attention, and where they occur, great presence of mind, and caution, are necessary to ascertain the cause, and manner of prevention, and so to apply a remedy that the life of the distiller may not be endangered; a wet sheet to throw over the doubling still, would always be proper.

PART II.

HAVING in the preceding pages given such directions as appeared necessary for the establishment of a grain distillery, and for conducting all the operations thereof, we now proceed to the consideration of the cider distillery and notice of some articles of the growth of the United States, which may be converted into spirit; a notice of the process of rectification necessary to the production of *alcohol* from grain and cider spirit, and to prepare a pure and neutral spirit for compound waters or the operations of the *cordial distillery*. Upon each of these subjects we shall offer such remarks as occur, when they come under consideration.

For the insertion of the other articles which will be found added, we shall offer no other apology than the general connection of some of them with the distilling business, and the evident advantage which must result to the distiller from a knowledge of the method of making all kinds of beverage.

Chapter I.

Of the situation for a Cider Distillery.

The situation for a common cider distillery, is not difficult to be fixed on; nothing further is necessary than a sufficient quantity and fall of water for the worm tub. If, however, an extensive establishment is contemplated, care should be taken to get into a good cider country; and in fixing the works, there will be great advantage in having the press in a situation so elevated that the cider should run from it to the fermenting vats, and from them, if possible, into the stills.

The vats should be under grounds, and of size sufficient to hold one day's pressing each; two days' pressing should not be put together.

Hogsheads will be found very troublesome and expensive for a large cider distillery.

Chapter II.

Of the various products of the United States which afford ardent spirits, by distillation.

The United States abound in fruits, roots, and vegetables, which will yield spirit upon distillation; an enumeration of these, however, with the various methods of treating them to the best advantage, would be as tedious as unnecessary.

For whilst our farmers raise such superabundant quantities of grain, and orchards continue to be cultivated, few other subjects will be worthy of the attention of the distiller, excepting merely for the gratification of his curiosity, or of his pride in displaying a variety of liquors.

To such as are actuated by these motives, we trust, that the remarks and directions contained in the work, will afford a sufficient clue to guide them in any experiment, they may wish to undertake; similar substances requiring similar modes of treatment.

I shall content myself with noticing such as appear generally useful.

Of Apple Brandy, or Cider Spirits.

The great quantities of apples raised in different parts of the United States, which cannot be disposed of in any other way, render them an object to the distiller. In many places, farmers are provided with stills merely for the distillation of their own fruit, and that of a few neighbours.

In the state of New Jersey, it has become so great an object, that large works are erected for the purpose, with stills of upwards of 1,000 gallons.

The custom generally at those works is, for the farmer to carry his apples to the distillery, where he receives one gallon of brandy for every five bushels. The apples after being assorted, so as to work the ripest first, are then ground, either in the common way, with nuts, or in a mill, constructed similar to the tanner's bark mill; after which it is pressed in a large powerful screw press, as long as any juice can be obtained. The cider is then put into large cisterns, or vats, prepared for the purpose; where it undergoes a fermentation, and is fit for the still, in from six to twelve days, according to the weather. Some distillers preserve the pomage after pressing, put it into casks, and cover it with water, until it undergoes a fermentation, when it is again pressed out, and the cider distilled. This however requires so much work and so many casks, that in a busy season it is not worth attending to. Throughout Lancaster county, and indeed in many other places, it is customary, after grinding the apples to throw them into casks, where they undergo a fermentation, after which the whole mass is committed to the still. This is very subject to empyreuma and the spirit obtained is of a very inferior quality; though it is said a greater quantity can be obtained in this than in any other way. From the tediousness of the operation, I am inclined to think it will be eventually found the least profitable, if an experiment were fairly and properly made. To judge of the progress of the fermentation, run a stick down in the centre of the cask; if upon drawing it out it is accompanied with a bubbling hissing noise, the fermentation is not over, but if no such noise is observable, it is then fit for the still. One of the great advantages stated in favour of steam stills is the distillation of pomage, by which a considerably greater quantity of spirits may be obtained.

To those who are desirous of following this plan, I would advise, as the best method of avoiding an empyreuma, the filling the still one third, or one half with water, which must be made to boil before putting in the pomage.

This is properly termed apple brandy, and the former, cider spirits; a distinction, which it is to be regretted is not more generally made, as it would give to cider spirits its just value in the public opinion. The two kinds however are too generally blended together at market, under the name of apple whiskey.

Of Peaches.

This fruit, which is equal if not superior in point of flavour to any in the world, grows abundantly in different parts of the United States, and yields upon distillation a spirit of remarkably fine flavour, principally valued, for the purpose of forming agreeable mixtures.

The method of treating peaches and apples are similar. By some, the fruit is thrown into a large trough, where it is pounded with large pestles until completely mashed; it is then pressed out, and a hogshead of pure peach juice obtained in this way, will yield from ten to twelve gallons of the best brandy; as the pomage cannot be completely pressed, it is then thrown into casks, diluted with water, and after sufficient fermentation again pressed, and immediately distilled.

Another method, and the best, where a large quantity of peaches are to be distilled, is to grind them with iron nuts; which by mashing the stone and kernel, is said to impart an agreeable bitter to the spirit; in this state it is fermented, and with the addition of a small quantity of water committed to the still. Others press it after the manner of pressing apples, which is preferable.

Of Cherries.

Cherries maybe treated in the same way as peaches and yield an agreeable spirit; the better method, however is to make what is called cherry-bounce or cherry-brandy, as follows: Fill a cask with cherries, sound and fresh picked, two thirds morello, and one third black cherries; then fill up the cask with brandy, or pure rectified whiskey; after standing about a month, it may be drawn off, and by the addition of a small quantity of sugar and spices, a fine cordial is obtained; or it may be used without this addition.

The cask must then be again filled with whiskey, and in a few weeks it will be fit for use; and though not so pleasant as the former, it affords an agreeable beverage when mixed with water. When this second filling is drawn off, the cask may be filled with water, which extracts all the spirit from the cherries.

Fox Grapes

Grow so abundantly in many parts of the United States as to render them an object worthy of the attention of the distiller. From their richness, there is little doubt but that a fine wine might be made from them, and a spirit little inferior to the best Cogniac brandy. Let them be mashed, and after standing two or three days, the juice must be expressed and fermented, as with peaches. The spirit produced upon distillation will be very good, of itself, and serviceable in imitating Cogniac brandy.

Concerning Potatoes.

Potatoes afford a good crop to the agriculturalist, yield a quantity of fine pure spirit, and afterwards are useful as food for hogs, cows, and sheep. It is a crop not always successful; but where a farmer can, as is frequently the case, raise upwards of 200 bushels from an acre of ground, there are few things will be found so profitable. Being well worthy the attention of the distiller, three recipes are given for extracting spirit from them. They are very different, but each may succeed. It is worthy of remark, that Dr. Anderson's experiment was made in the *spring* of the year, when it is probable the potatoes were somewhat *sprouted*. His plan has been tried in the month of August, without success, but this may have been owing to other causes.

The steam of the *boiler* in a distillery might be used to boil them, without any expense, other than keeping up the boiling heat for the necessary time, which will be trifling.

*Method of extracting Spirit from Potatoes, practised by Mons. Bertrand, at Metz.**

Take 600 pounds of potatoes, and boil them in steam about three quarters of an hour; till they will fall to pieces on being touched. The vessel in which they are boiled consists of a tub, somewhat inclined; in the lower part of it are two holes, one for the purpose of bringing in the steam produced in another vessel, over a coal fire, and the other made to carry off occasionally the condensed water. After the potatoes are boiled, they are crushed and diluted with hot water till they are of a liquid consistence; then add 25 pounds of ground malt, and two quarts of yeast; the mixture is to be stirred, covered with a cloth, and kept to the temperature of 15° Reaumur, or 66° Fahrenheit.

*The *Countess de N.....* near Vienna, has a distillery of potatoes. This is rather a singular occupation for a *titled lady*. The process appears to be very correct.

After fermentation, and the exhalation of the carbonic acid, the matter sinks down and is fit for distillation; by means of two stills, this mass may be rectified in one day, and it will produce about 44 quarts of spirit, worth a guinea and a half, while the whole cost, including coals and labour, is about 23 shillings and six pence. The residuum is good food for hogs.

Method of preparing Potatoes for Distilling.

(By Samuel M[c]Harry, Esq.)

"Wash them clean, and grind them in an apple mill, and if there be no apple mill convenient, they may be scalded and then pounded; then put two or three bushels into a hogshead and fill the hogshead nearly full of boiling water, and stir it well for half an hour; then cover it close until the potatoes are scalded quite soft; then stir them often until they are quite cold; then put into each hogshead about two quarts of good yeast, and let them ferment, which will require eight or ten days; the beer then may be drawn off, and distilled, or put the pulp and all into the still as you do apples. I have known potatoes distilled in this way to yield upward of three gallons to the bushel."

Another Method.

Dr. Anderson of Scotland gives an account of a very fine spirit which he procured from potatoes; he says, "It was somewhat like very fine brandy, but milder, and had a kind of coolness on the palate peculiar to itself. Its flavour was still more peculiar, and resembled brandy impregnated with the odour of violets and raspberries."

In February, he boiled to a soft pulpy state, a bushel of them, weighing seventy-two pounds; then bruised and passed them through a straight riddle along with spring water, keeping the skins back in the riddle and throwing them away. Cold water was added to the pulp and mixed up till the whole mixture was twenty gallons. It stood until sufficiently cool, when yeast was mixed with it as if it was malt wort.

In 10 or 12 hours a fermentation began, which continued very briskly 10 or 12 hours; and then began sensibly to abate. It was now *briskly stirred*, and the fermentation was thereby renewed. The same operation was renewed every day, and the fermentation thus continued for two weeks. It could not then be further kept up. It was now distilled, taking care to *stir* it to prevent empyreuma.

Of Beets.

On the subject of making sugar from beets, various experiments have been made in Prussia, France, and Austria, for some years past; and so confident was the Emperor Napoleon of success, that he ordered the planting of 32,000 acres of land in beets, for the purpose of supplying the necessary quantity of sugar, and prohibited the importation of that article, after the first of January, 1812.

It appears by calculation, that 300 acres will produce 133,200 kilogrames (266,400 pounds, American) of crude sugar, which will not loose more than one-eighth in refining. As this may become of importance to the American farmer and distiller, and the process for making spirit is the same, I have introduced the following:—

Account of the process used by Mr. Achard, for extracting Sugar from Beet-roots.

(From the *Annales de Chemiè.*)

The species of beet proper for making sugar is the *Beta vulgaris* of Linnæus; but all the varieties of that species of root are not all equally proper.

That of which the inside is white, the skin pale red, and the root long and spindle shaped, is the best.

Boil the root (with the skin, as it is taken out of the ground, and without any other preparation than that of carefully taking away the leaves and heart) in water till it is so soft that it may be penetrated by a straw; a short boiling is sufficient to produce this degree of softness, which is very well known to confectioners, and is given to several sorts of fruits before they are preserved.

The beet root, after cooling, is divided, and cut into slices, by means of a machine made use of by husbandmen for dividing potatoes for the use of cattle. This machine is described, in Burch's publication, entitled, *Ucbersicht der fortschritte in wissenchaften, kunsten, manufacturer, und handeverken; von Ostern 1799, bis ostern 1797, Erfurt 1798.*

Two men, with the assistance of the machine can cut nearly 100 pounds of roots in very thin slices in three minutes.

In order to extract the juice from the roots, after being sliced, they are submitted to the action of a press, which ought to act very strongly, that as much juice as possible may be drawn from them. The pulp remaining in the press still contains a considerable portion of sugar, to extract which, add a sufficient quantity of water, and after twelve hours press it out. After the second extraction, there still remains in the pulp a sufficient quantity of saccharine matter, to furnish advantageously, by means of fermentation, either brandy or vinegar. Mr. Achard states that he has on his estate in Silesia, a manufactory capable of furnishing 400 pounds of sugar per diem, for six months.

Concerning Persimons.

Several years ago, Mr. Isaac Bartram was requested by the American Philosophical Society to make some experiments of the distillation of persimons. The lateness of the season prevented him from making more than one trial, which was done with half a bushel of the fruit, in the month of December, when it was much damaged by the frost and rain. The success of this experiment, however, was such as to leave no doubt but that it was a matter welt worthy the attention of the farmer and distiller, and he recommends the following process:

Let a number of empty hogsheads, in proportion to the quantity of fruit, be provided; take out one of the heads of each, and in the other let a hole be bored, at about four inches from the chimb, into which fix a ping, which may be occasionally taken out from the lower end, when the casks are fixed upon trussels, at a small distance from the ground. In these casks, over the holes, lay a number of small sticks, covered with straw, about two or three inches thick, to prevent the pulp from choaking them.

Four hogsheads being thus prepared, fill one of them half full with persimons, which have been well mashed; add water until it arise within one third of the top; then cover the cask with the head that has been taken out, and let it stand about nine days; by this time the pulpy or feculent part of the fruit will be separated by the act of fermentation; you are then to draw off the liquor, by the hole in the bottom of the hogshead, and put it in a tight cask, closely bunged up, to prevent a second fermentation, whereby your liquor would become acid, and be rendered unfit for the still.

Having thus extracted the more vinous parts from the first hogshead, let as much water be added as before, which must be well stirred, and mixed with the pulp, thereby to procure the whole strength of the fruit.

A second hogshead is then to be charged half full of fruit, well mashed as the first, and instead of pure water, fill it two thirds full with the second extract of the first hogshead, leaving it to ferment as before directed. This fermentation being perfected, draw off the liquor and let it be bunged up close. The third hogshead is to be treated as the second, and in the like manner every succeeding cask. After you have, in this manner, converted all your fruit into a fermented liquor, let it be kept at least one month before it is distilled, if it can be preserved without danger of becoming sour; for I have observed that vinous spirits drawn from new fermented liquors are not equal in flavour to those which have been meliorated by age.

Chapter III.

Of Colouring Liquors.

This is a part of the business of no use to the manufacturer of merely gin and whiskey. Those however who make brandy and spirits, will *colour* with brown sugar highly burnt or boiled in an iron pot over the fire; or they will probably find it easier and cheaper to purchase *colouring matter*, which may be had in all cities; the quantity necessary, depends upon the quality; here the person using it, must act according to his own discretion.

A *double handful* of parched, or burnt wheat, will give an agreeable colour to a *barrel* of whiskey, and will improve the flavour.

A quantity of oak shavings digested for some time in spirits of wine, will form a dilute tincture of oak, which is in reality the cause of the colour of French brandy; this may be added to colour spirits instead of burnt sugar.

Chapter IV.

*Of Rectification.**

The prime object of rectification is to free the spirit from the essential oil of the ingredients, from which it has been distilled; or from any disagreeable flavour, which it may have received. In order that this may be more easily accomplished, very great care should be taken in the first distillation of the spirit; that is, that it should be run off with a slow fire, by this means but a small quantity of the oil comes over with the spirit, and that not so intimately blended, as by a rapid distillation. This, though by no means a complete rectifier, has the effect of rendering the spirit much more mild, and easier to be operated upon, by a future process; for it has been made very evident, that it is much easier, to prevent the mixture of a great part of the essential oil, than to free a spirit from it, which has become completely impregnated.

A great variety of methods have been tried, to effect this desirable object, though without success; and there is scarcely a distiller without his favourite nostrum; these in general are so inefficient, and many so absurd, that a bare enumeration of them, would be a waste of time and paper; suffice it to say, that where they succeed in taking off the objectionable flavour of the spirit, they generally leave a worse.

Although no efficient method has been made public† for freeing spirit from a disagreeable flavour by distillation, yet we are in possession of a way of effecting this

*In treating of rectification, the author has avoided all those tedious, expensive, and ridiculous processes made use of in this country and in England, previous to the discovery of the effect of charcoal, on spirits; and in this and the following chapter, he has endeavoured to give only such hints, as may be certainly useful to the American distiller; and mentioned ingredients which are attainable at a reasonable expense.

†When I was first made acquainted with Mr. Allisons's process for neutralizing spirit, I thought the effect could be more speedily obtained by distillation. It was, however, considered as the peculiar excellence of his plan, that the expense of a still or of fuel was unnecessary, I thought no more of the subject, therefore, until some time in the year 1806 or 1807, when I happened to have some *still burnt* whiskey, and, at the same time, lying by me, some charcoal, which was not good enough to use in filtration. As my whiskey could not be made worse, I determined to take this opportunity of making an experiment, and accordingly threw into the still, a quantity of each, and agreeably to my expectation, produced a spirit free from the empyreumatic flavour, though not perfectly pure.

I have repeated the experiment since, and am now convinced, that a perfectly pure spirit may be obtained by distillation of any whiskey, however impure, with fresh coal; and that charcoal not good enough for filtration may yet be used with advantage in the above mentioned manner.

Here was a *discovery* for which a patent, no doubt, might have been obtained; but I thought it too nearly allied in principle to Mr. Allison's plan, and did not prosecute my right. But I have since heard, with some desirable object, which, though somewhat tedious, is very complete, and leaves the spirit perfectly pure, and entirely divested of the aroma or essential oil.

For this important and very valuable discovery we are indebted to the scientific researches of the Rev. Burgiss Allison, now of Washington, D.C. who as early as 1786, made the discovery, which he then communicated to several of his friends.

In 1802, he obtained a patent for improving spirits, and in 1803 made an "improvement in the application of the principle of rectifying or improving spirituous liquors."

Through the politeness of Dr. A. I have been allowed the liberty of publishing his process, at the same time cautioning the public against using it without the liberty of the patentee, by application to whom, however, rights may be obtained on very liberal terms.

Process.

Procure a quantity of good maple or chesnut charcoal, taking care to get such as has not been exposed to the rain or heavy dews; let this be ground perfectly fine, and at all times kept as dry as possible. Next, get a proper kind of vessel, say a half hogshead, in which must be fitted very nicely a second or false bottom, about four inches from the other, perforated with as many holes as can conveniently be made with a very large gimblet; a hole must then be made between the two, or in the lower bottom, for the purpose of drawing off the liquor as rectified.

The cask must now be placed in a firm position so that a barrel will stand under it, to receive the liquor. Two pieces of flannel, cut to fit very exactly, must now be laid on the false bottom. Then in another tub mix, or rather moisten well, a quantity of charcoal, with the liquor to be rectified. Strew this paste or mixture closely over the flannel, to the thickness of about an inch, so that no crack or crevice is left; it is then ready to receive the liquor to be rectified; but to avoid disturbing this paste, by pouring on the liquor, it will be advisable to cover it with a piece of gause, and also to put into the tub, a small piece of board on which the liquor should be gently poured.

This tub so prepared, is now capable of rectifying 300 gallons without being removed; to do which proceed as follows:

In another tub to be placed close along side of this, mix with a quantity of the liquor to be rectified, as much charcoal as is necessary to deprive it of its peculiar flavour, which will be about one eighth, according however to the quality of the charcoal, and the strength of the essential oil or flavour to be destroyed; after standing a few minutes, this is now to be gently poured into the filtering tub until it is full; the liquor will soon run through, and after the first quart, will be found perfectly pure and tasteless. By pouring on the liquor too fast at first, it will sometimes get down the sides of the tub, and by blacking the lower part of the cask, render it unfit for the operation, until

surprise, that a Mr. Parsons, has obtained a patent for a similar plan. How far we may differ, I cannot say; as I have been disappointed in my hope of obtaining a specification of his patent. It is however proper to mention it, as a caution to those who are in the habit of using charcoal, least they may thereby interfere with the right of another.

cleaned; to avoid this, it will be better to pour in about four gallons of liquor, mixed with a larger proportion of charcoal than necessary; when this runs perfectly pure, the cask may be filled without danger of accidents, twice every day.

After the process has been continued some days, the cask becomes nearly full of charcoal, and cannot be longer used, until emptied; but this charcoal has retained a quantity of spirit; to extract which, water must be poured on above, so long as any spirit remains; a part of the spirit runs out of equal strength, with what was used; it, however, gradually becomes weaker, until there is nothing but water; the weaker part must be distilled.

It may be proper to observe that nothing should be used in this operation, that can possibly give a flavour to the spirit.

As it is sometimes difficult to get charcoal ground at a mill; a writer has recommended the following method; he says, "I have constructed a small hand mill, with which, one person may grind or pulverize, sufficient for rectifying 200 gallons of spirit per day. This is done by means of a screw, made precisely in the form of a common screw augur, about 18 inches in length 5 inches in diameter, gently tapering to the extremity, to about 1 inch. This screw is laid in a wooden bed, into which the large lumps are thrown; and, by turning this by hand, the large lumps are by that means crushed. After passing the length of the screw, it is found broken into pieces, of the size of grains of Indian corn, and falls into the hopper of a small corn mill, in the form of an iron coffee, mill, though larger, and is worked by the same power."

"By this means the process of pulverizing is greatly facilitated, and one man may, as above stated, furnish any desired quantity with convenience, as he may enclose all his works in a box or linen, and deliver it into another closed vessel, without making any dust, and save his lungs from the disagreeable effect created in operation."

Chapter V.

Concerning the Imitation of Foreign Spirits.

By the process of rectification detailed in the preceding chapter, the operator will be in possession of a perfectly pure and tasteless spirit; to which any flavour that may be desired, can be easily given by the application of the proper ingredient, or use of the essential oil.

Gin, brandy, and Jamaica rum, have been so well imitated as to deceive very good judges, but so many unsuccessful attempts have been made by unskilful and injudicious operators, that a prejudice has arisen against what is termed patent brandy, which is heightened by the general dislike of the mixture and adulteration of spirits.

The usual method is to mix one gallon of the brandy or spirit, to be imitated, with two gallons of rectified or neutralized spirit; which is its most appropriate name; the proper proportions, however, must depend upon the purity of the neutralized spirit, and the relative flavour of the brandy, or other spirit which may be used; as the greater the quantity of the essential oil it may posses, the smaller proportion will effectuate the purpose; much then depends upon the quality of the ingredients, and it is requisite that the operator be a man of correct taste to be able properly to apportion them. But as all spirit is radically the same, receiving its peculiar flavour from the presence of an essential oil, a certain portion of which is necessary, it is evident that no attempt at imitation can be completely successful, without having this due portion of essential oil.

Thus, although a mixture of one gallon of French brandy, and two gallons neutralized spirit will *smell* exactly like brandy, yet will there be deficiency of two third parts of essential oil, which will, however be only detected when mixed with water, and by one accustomed to the full, luscious taste given by the essential oil of Bordeaux, or Cogniac brandy. An ingredient, therefore, possessing the flavour of brandy is here wanted to supply this deficiency; none such has as yet been discovered in this country.

It is obtained in England by fermenting dried wine lees and extracting therefrom a spirit strongly impregnated with the *essential oil*, of which a sufficient quantity is added to the neutralized spirit to give the desired flavour.

These lees may be imported from France; but the American distiller who makes the attempt must be careful, to have the *kind* designated, lest he may through ignorance or mistake, endeavour to make Cogniac brandy, from Bordeaux lees.

A spirit also may be obtained by fermenting raisins with water, and a small quantity of sugar, which will be highly serviceable.*

Another method is to scorch or partially burn a quantity of Prunes, and infuse them in the neutralized spirit. They impart a rich luscious flavour, and the addition of about one-eighth part of strongly flavoured brandy renders the imitation very complete. This plan is objectionable on account of the cost of prunes and the trouble of preparing them.

The quantity necessary can alone be determined by experiment.

Jamaica, and other kinds of rum generally contain so large a portion of essential oil, that they are rather *improved* by a mixture with an equal portion of neutralized spirit.

To give an agreeable vinosity to this brandy a few drops of sweet spirits of nitre may be added.

*Dr. Clark, in his travels, relates, that about 150 or 200 vessels are employed to carry *nardec*, a marmalade of grapes, and *becmis*, a syrup made from various fruits by boiling them with honey, from Trebizond and Sinope, to Tagenroy in Tartary. Raisins of the sun, are also taken in great quantities. All these are used in the distilleries; and the spirit is sold through the Russian empire as French brandy.—page 267.

Chapter VI.

Of Alcohol, or Spirits of Wine.

The term alcohol is applied, exclusively, by modern chemists, to the purely spirituous parts of all liquors that have undergone the vinous fermentation.

Alcohol is in all cases, the product of fermentable matter, and is formed by the successive processes of vinous fermentation, and distillation. All fermented liquors, therefore, agree in these two points; the one, that a saccharine juice has been necessary to their production; and the other, that they are capable of furnishing an ardent spirit by distillation.

Various kinds of ardent spirits are known, in commerce, such as brandy, rum, arrack, malt spirit, whiskey; these differ from each other in colour, smell, taste and strength; but the spirituous part to which they owe their inflammability, their hot fiery taste, and their intoxicating power is the same in each, and may be procured in its purest state by rectification, which is performed by means of charcoal, and by repeated distillation.

Alcohol, as well as ardent spirits of different kinds, is procured most largely in this country, from a fermented grain liquor, prepared for the express purpose of distillation from grain; but in the wine countries, the spirit is obtained from the distillation of wine; hence the synonimous term, spirit of wine.

Alcohol is a colourless transparent liquor, appearing to the eye like pure water. It possesses a peculiar penetrating smell, distinct from the proper odour of the distilled spirit from which it has been procured. To the taste it is excessively hot and burning, but without any peculiar flavour. From its great lightness and mobility, the bubbles which are formed on shaking it, subside almost instantaneously, and this is one method of judging of its purity. Alcohol is very easily volatilized by the heat of the hand; it even begins to be converted into vapour at the lowest temperatures, absorbing heat from surrounding bodies, as it assumes that elastic form. It boils at about 176, and the vapours when condensed, return unaltered to their former state. It has never been frozen by any cold,* natural or artificial, and hence its use in thermometers to measure very low temperatures.

*Since this was written, Mr. Hutton, of Edinburgh, has announced some experiments upon the freezing of alcohol, which requires a degree of cold that has never before been produced by any means, and which is stated by him to be 116 degrees below zero of Fahrenheit. The fluid was frozen to a perfectly solid mass, composed of three strata, the uppermost of a yellowish green, the second of a pale yellow colour; and the

Alcohol takes fire very readily upon the application of flame, the speedier in proportion to its purity. It burns with a pale flame, white in the centre and blue at the edges; gives but a small degree of heat, and is so faint as to be scarcely visible in broad day light. It burns without any smoke or vapour, and if strong leaves no residuum; but if weak, is extinguished spontaneously, and the watery part remains behind.

Alcohol mixes with water in every proportion. Heat is extricated during the mixture, which is sensible to the hand, even in small quantities. At the same time there is a mutual penetration or concentration of parts, so that the bulk of the two liquors, when mixed, is less than when separate, consequently the specific gravity of the mixture is greater than the mean specific gravity of the two liquors taken apart. The alcohol may be again for the most part separated from the water by distillation with a gentle heat. Thus, if one part of alcohol, whose specific gravity is 817, be mixed with one part of water, whose specific gravity is 1,000, the specific gravity of the compound will not be the mean of their respective gravities, 908.5, but will not be less than 934; and a difference will likewise obtain in whatever proportions the fluids are mixed.

This difference proceeds in a decreasing ratio; that is to say, when ten parts of alcohol are mixed with one of water, the difference in the specific gravity, which is produced, is greater than when the ten parts are mixed with two, and greater in this than when mixed with three.

The progression, however, is not regular, and hence the specific gravity of any possible mixture of alcohol and water, cannot be calculated *a priori*, but must be determined by actual experiment.

When alcohol is diluted with an equal weight of water, it forms what is called proof spirit, but has not the same flavour with distilled spirit of the same strength, in which probably the combination is more intimate, and it is known indeed, that the increase of density from such a mixture, does not at once obtain its maximum, proving, therefore, that the combination is not at once complete. Hence the impropriety of laying by very *strong spirit* to be meliorated by age; it should not be more than second or third proof.

Alcohol is capable of uniting with a great number of substances, a circumstance which renders its use very extensive in a variety of chemical processes, and in analysis.

A peculiar colour is perceived in the flame of the solution of some of the acids in alcohol, when set on fire. The solution of nitre gives a pale yellow flame, that of boracic acid is a faint green, all the solutions of copper burn with a beautiful bright green, and those of nitrated or muriated strontian, shine with a deep blood red.

Alcohol is an excellent solvent for some essential oils, and in general for the most odorous and inflammable of the vegetable productions.

In the essential oil of a plant resides the *spiritus rector*, or the aroma, that which gives the exquisite perfume to the rose or jessamine; when these odoriferous plants are

third which greatly exceeded the rest in quantity, and was the pure alcohol, nearly transparent, and colourless. It was proved that the alcohol was not decomposed in the process, but merely separated from two foreign substances which it had held in solution; these are highly volatile, and cannot be separated but by freezing; to them the alcohol owes its peculiar flavour. Mr. Hutton has not made public the method of producing this degree of artificial cold. See *Repertory of Arts*. vol. 34.

distilled with alcohol, it rises strongly impregnated with their scent and flavour, and as it takes up no colouring matter it remains perfectly clear as before. Thus, the common lavender water is alcohol distilled off with the lavender plant, and holding in solution the essential oil in which the scent resides. The distilled spirits in pharmacy, are similar preparations of alcohol, containing the flavour of spices, aromatics, or other substances with which it has been distilled.

Various tests have been devised to ascertain the purity of alcohol, and the proportion of water which it contains. A spirit, which is very free from water, will, when set fire to, burn away without leaving any residue; if it is of moderate strength it will burn for a certain time, and then become extinguished, and leave a portion of water, more or less considerable, according to the degree of dephlegmation; if on the contrary, it is very watery, it will not kindle at all.

This test, however, is by no means accurate, since the heat of the burning spirit will evaporate part of the water which should be left in the residuum.

Another test, is to pour a small quantity of spirit on a small heap of gunpowder and kindle it. The spirit burns quietly on the surface of the powder until it is all consumed, and the last portion fires the powder if the spirit was pure, but if watery, the powder becomes too damp and will not explode. This test, also, is very inaccurate, for if the powder be drenched even with a strong spirit, it remains too damp to be fired; and, if it be only barely moistened, any spirit that will burn will inflame it. A better test is, to shake the spirit in a vial with some dry carbonated alkali; pearl or pot ash recently dried by a strong heat; but the most accurate of all is to ascertain its specific gravity, and compare it with the density of known quantities of alcohol and water, previously mixed for the purpose of giving a standard of comparison, and this is best determined by the hydrometer or gravimeter.

Chapter VII.

Of Bodies proper for Distillation, and their products.

The proper subjects for distillation are flowers, fruits, seeds, spices, and aromatic plants.

By distillation, and digestion, we extract the colour and smell of flowers in simple water, and essences.

We extract from fruits, at least from some, colour, taste, &c.

From aromatic plants, the distillers draw spirits, essences, simple, and compound waters.

From spices, are produced essences, or in the languages of chemists, oils and perfumes, and also pure spirits.

From seeds and berries are drawn simple waters, pure spirits; and from some, as those of annise, fennel, and juniper oil.

The colour of flowers is extracted by infusion, and likewise by digestion, in brandy or spirits of wine; the smell is extracted by distillation; the simple water with brandy, or spirits of wine.

Substances are said to be in digestion, when they are infused in a menstruum, over a very slow fire. This preparation is often necessary in distillation, for it tends to open the bodies, and thereby free the spirits from their confinements, whereby they are better enabled to ascend. Cold digestions are the best; those made by fire, or in hot materials, diminish the quality of the goods; as some part, as the most volatile, will be lost. It is of absolute necessity for extracting the spirits and essences of spices.

In bodies that have been digested, the spirits ascend first; whereas in charges not digested, the phlegm ascends before the spirits.

Another remark should be mentioned. That in mixed charges, consisting of flowers, fruits, and aromatic plants, put into the alembic without previous digestion, the spirits of the flowers ascend first; and notwithstanding the mixture, they contract nothing of the smell, or taste of the fruits, and plants. Next, after the spirits of the flowers, those of the fruits ascend, not in the least impregnated with the smell or taste of either the flowers or plants. And in the last place, the spirits of the plants distil no less neatly than the former. Should this appear strange to any one experience will convince him of the truth.

Chapter VIII.

Of Distilling Simple Waters.

The plants designed for this operation are to be gathered when their leaves are at full growth, and a little before the flowers appear, or at least before the seed comes on; because the virtue of the simple expected in these waters is often little, after the seed or fruit is formed; at which time plants begin to languish. The morning is the proper time to gather them, because the volatile parts are then condensed by the coldness of the night, and kept in by the dew not yet exhaled by the sun.

Flowers should be gathered before they are quite opened; and seeds, when they arrive at perfect maturity.

When the desirable property is in the root, it should be dug in the winter or spring while full of sap. And the bark or sap is most strongly impregnated during the summer.

The subject being chosen, let it be bruised, or cut if necessary, and with it fill two thirds of a still, leaving a third part of it empty, without squeezing the matter close; then pour in as much rain, river, or spring water, as will fill the still the same height; then fit on the head, and let the plant digest, with a small degree of heat, as long as may be thought necessary; after which raise the fire and distil as usual.

Such is the general method of procuring simple waters; the following rules however are necessary to render it applicable to all sorts of plants.

1. Let the aromatic, balsamic, oily, and strong smelling plants, which long retain their natural fragrance, such as balm, hyssop, juniper, marjoram, pennyroyal, mint, rosemary, lavender, sage, &c. be gently dried a little in the shade; then digest them for twenty-four hours, in a close vessel, and afterwards distill, and they will afford excellent waters.

2. When waters are to be drawn from seeds, barks, or woods, that are very dense, ponderous, tough, and resinous, let them be digested for three, four or more weeks, with a greater degree of heat, in a close vessel, with a proper quantity of salt added, to open, and prepare them better for distillation, and prevent putrefaction, which would certainly happen without it.

3. Those plants which diffuse their odour, to some distance from them, and thus soon loose it, should be distilled immediately upon gathering in the proper season,

without any previous digestion; thus borage bugloss, jessamine, white lilies, lilies of the valley, roses, sweet briar, lilac, &c. are hurt by heat, digestion, or lying in the air. They may, however, be preserved by being packed in close vessels with salt.

Lastly, those that contain a more fixed oil, should be imperfectly fermented, by which process we obtain the virtue very little altered from its natural state, though rendered much more penetrating and volatile. The operation is performed in the following manner:

Take a sufficient quantity of any fresh plant, cut it, and bruise it, if necessary; put it into a cask, leaving a space, empty at top, of about four inches, then take as much water as would, when added, fill the cask to the same height, including the plant, and mix therein about an eighth part of honey, if it be cold weather, or one twelfth, if it be warm, or the same quantity of sugar will do, or half an ounce of yeast to each pint of water will have the same effect. When the proper quantity of honey is added to the water, let it be warmed, and poured into the cask, and set in a warm place, to ferment for two or three days; but the herbs must not be suffered to fall to the bottom, nor the fermentation be more than half finished.

The whole must then be committed to the still and the fire raised by degrees; for the liquor containing much fermenting spirit, easily rarefies with the fire, froths, swells, and therefore, becomes very subject to boil over; we ought, therefore, to work slower, especially at first.

Thus may simple waters be made fit for long keeping without spoiling; the proportion of inflammable spirit generated in the fermentation, serving excellently to preserve them.

Chapter IX.

Of making Compound Waters and Cordials.

Mr. Cooper has given the following general rules as necessary to be observed in this branch of distillery.

1. The artist must always be careful to use a well cleansed spirit, or one freed from its own essential oil, for as a compound water is nothing more than a spirit impregnated with the essential oil of the ingredients, it is necessary that the spirits should have deposited its own.

2. Let the time of previous digestion be proportioned to the tenacity of the ingredients, or the ponderosity of their oil.

Thus cloves and cinnamon require a longer digestion before they are distilled, than calamus aromaticus, or orange peel. Sometimes cohobation is necessary, that is, returning the distilled water upon a fresh portion of the plant, &c. for instance, in making the strong cinnamon water, because the essential oil of cinnamon is so extremely ponderous, that it is difficult to bring it over the helm with the spirit without cohobation.

3. Let the strength of the fire be proportioned to the ponderosity of the oil intended to be raised with the spirit, thus for instance, the strong cinnamon water requires a much greater degree of fire than that from lax vegetables, as mint, balm, &c.

4. Let only a due proportion of the finest parts of the essential oil be united with the spirit; the grosser and less fragrant parts of the oil not giving the spirit so agreeable a flavour, and at the same time renders it thick and unsightly. This may in a great measure be effected by leaving out the feints, and making up to proof with fine soft water in their stead.

These four rules carefully observed will render this part of distillation easy and simple. Nor will there be any occasion for the use of burnt allum, white of eggs, isinglass, &c. to fine down cordial waters; for they will presently be fine, sweet and pleasant tasted, without any further trouble.

Recipe 1.—Clove Water.

Take four pounds of bruised cloves, half a pound of pimento, or allspice, and sixteen gallons proof spirits. Digest the mixture in a gentle heat, and then draw off fifteen gallons, with a moderately brisk fire. This as well as any other water, may be coloured with a strong tincture of cochineal, or colouring matter; and sweetened at pleasure with double refined sugar.

Recipe 2.—Lemon Water.

Take of dried lemon peel, four pounds; pure proof spirit, ten and a half gallons; and one of water; draw off ten gallons by a gentle fire; and dulcify as above.

Recipe 3.—Citron Water.

Take of the dry yellow rinds of citron, three pounds; of orange peel, two pounds; bruised nutmegs, three quarters of a pound; clean proof spirit, ten and a half gallons; water, one gallon. Digest them in a moderate heat; then draw off ten gallons, and dulcify as above.

Recipe 4.—Orange Water.

Take of the yellow part of fresh orange peel, five pounds; clean proof spirit, ten and a half gallons; water, two gallons; and draw off ten over a slow fire.

Recipe 5.—Lavender Water.

Digest in ten gallons rectified spirits of wine and one gallon of water, 14 pounds of lavender flowers; then draw off ten gallons for use; this should be done in Balneo mariæ, or water bath.

Recipe 6.—Lavender Compound,

May be made without distilling, as follows: Fill a gallon jug with lavender flowers; then pour on as much French brandy or pure rectified spirits, as the jug will hold, cork it up, and set it in the sun shine, shaking it daily; in a month or two it will be fit for use, when it should be poured off and the jug refilled.

Recipe 7.—Peppermint Water.

Digest in ten and a half gallons of proof spirits, and one gallon of water, 14 pounds of dry peppermint leaves; then with a moderate heat draw off 10 gallons, and dulcify at pleasure.

Recipe 8.—Compound Gentian Water.

Infuse in six quarts of proof spirits, and one quart of water, eight ounces of the leaves and flowers of the lesser centaury, three pounds of gentian root sliced, and six ounces of orange peel; then draw off until the feints begin to rise. A powerful preventative against fever and ague.

Recipe 9.—Anniseed Water.

Take three ounces of carroway seeds, six ounces of anniseeds, water one gallon, proof spirits four gallons; infuse them all night in your still, and draw off until proof with a slow fire; dulcify with white sugar at pleasure.

Recipe 10.—Another Way.

Distil in one gallon water, and twelve and a half gallons of proof spirits, two pounds of bruised anniseeds; then draw off 10 gallons with a moderate fire; and dulcify with white sugar. This is the method of making the Malta anniseed, esteemed the finest in the world.

This cordial is used by all classes of people in the United States, and may be made for any price by using low price spirits and dulcifying with brown sugar.

Recipe 11.—To make ten gallons Royal Usquebaugh.

Take of cinnamon, ginger, and coriander seeds, of each three ounces; nutmegs four ounces and a half; mace, cloves and cubebs, of each one ounce and a half. Bruise these ingredients, and put them into an alembic with 11 gallons of proof spirit, and two gallons of water; and distil till the feints begin to rise, fastening four ounces and a half of English saffron, tied in a cloth, to the end of the worm, that the liquor may run through it, and extract all its virtues. Take raisins, stoned, four pounds and a half; dates, three pounds; liquorice root, sliced, two pounds; digest these 12 hours, in two gallons of water; strain out the clear liquor; add to it what was obtained by distillation, and dulcify the whole with fine sugar.

Recipe 12.—To make Red Ratafia.

Take of black heart cherries, twenty-four pounds; black cherries, four pounds; raspberries and strawberries, of each three pounds; pick these fruits from their stalks, and bruise them; in which condition let them continue twelve hours; press out the juice, and to every pint of it add a quarter of a pound of sugar. When the sugar is dissolved, run the whole through the filtrating bag, and add to it three quarts of clean proof spirits. Then take of cinnamon, four ounces; of mace an ounce; and of cloves two drachms. Bruise these spices, put them into an alembic, with a gallon of clean proof spirits, and two quarts of water; and draw off a gallon with a brisk fire. Add as much of this spicy spirit to your ratafia as will render it agreeable to your palate; about one fourth is the usual proportion.

Ratafia made according to the above recipe, will be of very rich flavour, and elegant colour.

Some, in making ratafia suffer the expressed juices of their fruits to ferment several days; by this means the vinosity of the ratafia is increased; but at the same time, the elegant flavour of the fruits is greatly diminished. Wherefore if the ratafia be desired stronger or more vinous, it may be done by adding more spirit to the expressed juices.

It is also a method of some, to tie the spices in a linen rag, and suspend them in the ratafia. If this method be taken, it will be necessary to augment the quantity of spirit, first added to the expressed juice.

Another method is to bruise the fruit and immediately pour the spirit on the pulp; after standing a few days, express the juice, filter and add sugar and spices as before. This way, however requires rather more spirit, as it cannot all be pressed out of the skins, and other parts of the fruit remaining after the juice is extracted.

Gooseberries, mulberries, and honey cherries, may also be used for making ratafia.

A fine ratafia may also be made from the expressed juice of peaches, treated as above; except as to the spices, which will destroy the fine flavour of the peach.

From this recipe it may be seen that the famous ratafia differs very little from our common spiced cherry bounce; indeed the juice of our cherries, may be made the foundation for cordials, with a judicious use of spices, equal to any of the celebrated foreign cordials.

It is thought unnecessary to give any more recipes as it will be essentially necessary for the complete cordial distiller to consult the public taste in his compound, to have a correct taste himself, and be well acquainted with the different strengths and qualities of drugs and spices; he can then vary his mixtures, and always present something new and agreeable.

Chapter X.

Concerning Wines.

The following paper on making wine is taken from the *Archives of Useful Knowledge*, No. 3, vol. I. It is given entire, as well to shew the manner of making wine, as to let our farmers see the advantage of devoting a small spot of ground to the cultivation of a vine which is either entirely neglected or suffered to grow in such situations, as not to produce fruit. Mr. Cooper is a practical man, who makes his experiments with care and attention. His observations therefore are particularly worthy of attention.

"In the year 1777, Joseph Cooper, Esq. of New-Jersey, noticed a native grape vine* in his neighbourhood, that covered a red cedar tree, so as to have the benefit of both sun and air, and the fruit on the south and south-west parts being unusually fine and ripening early, he was induced to plant a cutting from it near his garden, where it grew for several years on a small arbour in a neglected state, bearing a few grapes of a good quality. He then pruned the vine, enlarged the arbour, and spread the vine thin and regularly on it, and secured it by tacking and tying, to prevent its being displaced by wind, which is very injurious to vines. The growth of the vine and the quality of the grapes soon exceeded his expectation, and induced him to enlarge the arbour to the size of sixty by forty feet, the whole of which the vine covered; he then extended his garden fence, so as to take it in, and manured under the vine by water from the barn-yard; and although the ground under the vine was covered with a strong sward of grass, which gave him three middling cuttings of grass, the vine produced the following crops of grapes.

"In the year 1807, it yielded thirty-six and a half bushels of grapes; three and a half of the best were eaten or given away; the remainder were pressed, and yielded 91 gallons of juice: to the pummice, a small quantity of water was added, and on being pressed, 26 gallons more of juice was obtained. Both parcels were made into wine, three bottles of which were presented to the Agricultural Society of Philadelphia, and found excellent. Some of it had been made with sugar, and some without.

"In 1808, the fruit was destroyed by rose-bugs and drought.

"In 1809, the vine yielded twenty-six and a half bushels of grapes, and made 85 gallons of juice; water was added as before to the pummice, and the liquor which then flowed upon pressure, was mixed with the first running. The wine was tart at first, but grew sweeter as it advanced in age.

*It is the *Vitis occidentalis* of Bartram, or Blue Bunch Grape.

"In 1810, it yielded forty-two and a half bushels of grapes, at one picking. Some had been previously taken off. A bushel of bunches weighed thirty-four pounds. Instead of water, Mr. Cooper added about twenty gallons of cider to the pummice, and mixed the produce of the first and second pressings; 150 gallons were thus obtained. Time only, can show the effect of this novel combination.

"One year he omitted water, and fermented the pure juice; but the next year owing probably to the quantity of tartar which it had deposited being re-dissolved, notwithstanding the cask had been well rinsed, and with gravel, after racking, it became tart, and he was induced to distil it for brandy, the quality of which was excellent. The addition of brandy to the wine when fermenting, increased the acidifying process. The wine was racked three times into a tub, but always returned to the same cask. If a fresh cask had been used, probably the acid fermentation would not have come on. But the same cask is preferable. Mr. C. thinks that if water be added, there will be no danger of a second fermentation from the deposition of tartar.

"A great advantage of the native species of grape in question over foreign grapes is, that the vine of the former is not injured by frost; whereas a slight frost destroys both the fruit and vine of the latter. Hence our native grape may be permitted to remain on the vines so late in the season, as that fermentation will not be affected by too great a heat. Mr. Cooper adds too, that they are not subject to blast or rot on the vines like foreign grapes.

"The vine covers a surface of sixty feet by forty, making 2,400 square feet; there are 43,560 square feet in an acre, and consequently an acre would admit eighteen arbours as large as Mr. Cooper's; but to allow free circulation of air, fifteen would be sufficient, and on this calculation Mr. Cooper concludes, that this number of vines, 'planted in a good soil and properly cultivated, would in five, or six years at farthest, cover an arbour as large as mine, and produce more and better fruit than mine does from one vine. And from the product of my single vine (which you have often seen), for several years past, I am confident that an acre of land, properly planted and cultivated with the best native grape vines that can be found within a few miles of almost any farm-house in New-Jersey, or perhaps any state in the Union, would produce grapes sufficient to make 1,500 gallons of wine annually in the way I have recommended. I need not mention its quality, as you have often tasted it.'"*

The following directions to make wine, by Mr. Cooper, contain his last improvements:

"I gather the grapes when fully ripe and dry, separate the rotten or unripe from the others, and press for distillation if the quantity is worth attending to; I then open the cider-mill so as not to mash the stems or seed of the grapes; then run them through, put the pummice or mashed grapes on some clean long straw, previously made damp, and laid on the cider-press floor, lap it in the straw, press it well, then take off the pummice and add some water, or I believe sweet unfermented cider would be better, and answer in lieu of sugar. After it has soaked awhile (but do not let it ferment in the pummice), press as before, put all together, and add sugar till it is an agreeable sweet. I have found a pound to a gallon sufficient for the sourest grapes, and white Havanna sugar the best; but sweet grapes makes the best wine without any sugar.

*Letter to the Editor, Dec. 8, 1810.

"I have heretofore recommended putting the sugar in after fermentation, but on experience find it not to keep as well, and am now convinced that all the saccharine matter for making wine should be incorporated before fermentation. Previously to fermentation, I place the casks three or four feet from the floor; as the filth works out, fill it up two or more times a day till it emits a clear froth, then check the fermentation gradually, by putting the bung on slack, and tighten it as the fermentation abates. When the fretting has nearly ceased, rack it off: for which purpose I have an instrument nearly in the shape of a wooden shovel with a gutter in the upper side of the handle; place it so as to prevent waste, and let it dribble into a tub slowly, which gives the fretting quality an opportunity to evaporate, tranquillizes the liquor, and hastens its maturity. When the cask is empty, rinse it with fine gravel to scour off the yeast that adheres to it from fermentation, then for each gallon of wine put in one pint of good high proof French or Apple brandy, fill the cask about one-third, then burn a sulphur match in it; when the match is burnt out stop the bung-hole, and shake it to incorporate the smoke and liquor; fill the cask, and place it as before, and in about a month rack it again as directed above; the gravel is unnecessary after the first racking. If the match should not burn well the first racking, repeat it; and if it don't taste strong enough to stand hot weather, add more brandy. I have racked my wine three or four times a year, and find it to help its ripening; have frequently had casks on tap for years, and always found the liquor to improve to the last drawing.

"Being fully of opinion that our common wine grapes are capable of producing wine as good and as palatable (prejudice aside), and far more wholesome than the wine generally imported at so great an expense: and a supply of that article being very uncertain, I am induced to urge the making wine of all the native grapes that can be procured; and in collecting them to notice the vines that produce grapes of the best quality, and which are the most productive, as this will enable persons to select the best vine to cultivate and to propagate from. This ought to be particularly attended to, as there are many vines which produce good grapes, but few in quantity, and others very productive but of bad quality: and I believe full half the number that come from the seed are males, and will never bear fruit. The sex is easily distinguished when in bloom, by the females showing the fruit in the heart of the blossom as soon as open, and the male presenting nothing of that kind.

"As the native grape-vine will not grow well from cuttings, the best way I know of to propagate them is by removing the vines, or laying branches in the earth to take root for a year or more, and when rooted remove them, or plant the seeds from the best kinds, and when in bloom dig up the males. If well cultivated, they will blow in three or four years, but will produce different kinds, the same as apples; and I have had some from the seeds superior to the parent."

Mr. Cooper observes in one of his publications:—

"In February or March, previously to the sap's running, I examine and trim the vines, observing which branches will suit best for training to different parts of the arbour, or whatever the vines are to cover; leaving a sufficiency of the strongest shoots to extend, or fill vacancies if wanted; then cut the other side shoots of the last year's growth that appear large enough for bearers, leaving not more than three or four buds or eyes and the diminutive ones; cut the dead and unnecessary old vines, close to the leading branches; then spread the vines regularly over what they are to run on, and secure them from being shifted, by tacking or tying.

"From trials and observations I am convinced, that the greatest error in making wine in our country is, using too much sugar and water for the quantity of fruit. The nearer wine is made from the juice of fruit, without water, the better, with no more sugar than will make it palatable by correcting the acid, and brandy or good cider spirit to give it strength sufficient to keep through our hot summers. The spirit will incorporate with the wine, so that when it arrives to proper age, it will not be known, by its taste, that any had been in it."

Recipe 1.—To make an excellent American Wine.

(Communicated by Joseph Cooper, Esq. of New Jersey.)

I put a quantity of the comb, from which the honey had been drained, into a tub, and added a barrel of cider, immediately from the press; this mixture was well stirred, and left for one night. It was then strained before a fermentation took place, and honey was added, until the strength of the liquor was sufficient to bear an egg. It was then put into a barrel; and after the fermentation commenced, the cask was filled every day, for three or four days, that the filth might work out of the bung-hole.

When the fermentation moderated, I put the bung in loosely, lest stopping it tight might cause the cask to burst. At the end of five or six weeks, the liquor was drawn off into a tub, and the whites of eight eggs, well beaten up, with a pint of clean sand, were put into it, I then added a gallon of cider spirits, and after mixing the whole well together, I returned it into the cask, which was well cleaned, bunged it tight, and placed it in a proper situation for racking off, when fine.

In the month of April, following, I drew it off into kegs, for use, and found it equal, in my opinion, to almost any foreign wine: in the opinion of many judges, it was superior.

This success has induced me to repeat the experiment for three years, and I am persuaded, that by using clean honey, instead of the comb, as above described, such an improvement might be made as to give a good wholesome wine, without foreign ingredients, at twenty-five cents per gallon, were every thing bought at market price.

Recipe 2.—Cider Wine.

The method of preparing this wine, consists in evaporating in a brewing copper, the fresh apple juice, until it be half consumed. The remainder is then immediately conveyed into a wooden cooler, and afterwards is put into a proper cask, with an addition of yeast, and fermented in the usual way.

Recipe 3.—To make Hydromel, or Mead.

To thirty gallons of water, add ninety pounds of pure honey, boil and skim it; put the liquor into a large open tub, and add two ounces of bruised ginger root, half an ounce of cinnamon, and the same quantity of pimento, or alspice; let the whole stand, until of a proper temperature, then add yeast, as in elder wine; flavour, and barrel it up for use, as directed for currant wine.

Recipe 4.—To make Currant Wine.

Take fourteen pounds of currants, when perfectly ripe, three gallons of cold water, break the currants into the water, and let them remain therein two or three days, and stir once a day. Strain the liquor from the fruit and stalks, and add fourteen pounds of sugar which being well mixed with *the currant* liquor, the whole may then be barrelled, and left fourteen days without the bung; after which bung it close, and bottle about Christmas, previously adding to every ten gallons one quart of brandy. The sugar should be of good quality, or honey may be used, adding about one-third more in weight.

If the flavour of orange peel (which is grateful in most wines of this description) is desired, a small quantity of the outer rind, will give it an agreeable flavour.

Sloes, bruised and infused in currant wine, impart a beautiful red colour, and a pleasant, rough, sub-acid taste, resembling that of *port wine.*

Recipe 5.—Elder Wine.

Take twelve and a half gallons of the juice of the ripe elder berry, and thirty-seven and a half gallons of water that has been recently boiled, and to every gallon of water, add three and a half pounds of sugar, or four and a half pounds of Havanna honey, which will incorporate whilst warm; add of ginger half an ounce, and pimento, three-fourths of an ounce to every four gallons of the mixture; and when the whole is cooled to about 60° Fahrenheit, add about half a pint of brewer's yeast, and let it ferment slowly, for about fourteen days, the bung being out; then bung it and let it stand six months, when it is fit to bottle.

Recipe 6.—Champagne Wine.

This wine has been imitated in England, with great success, by using gooseberries before they ripen, and supplying the want of saccharine matter, with loaf sugar.

In the province of Champagne, sugar is frequently added to the grapes, when they do not attain their maturity, for the preparation of the Champagne wine. Much of the wine which they export, is made in this way,

The imitation of it, with green gooseberries and sugar, is as salutary, very palatable and attainable in this country.

Recipe 7.—To make Irish Nectar.

The nectar of the Irish, was composed of honey, wine, ginger, pepper, and cinnamon. The French poets of the thirteenth century spoke of it with rapture, as being most delicious. They regarded as the very perfection of human ingenuity, the union of the juice and spirit of the grape, with the perfume of foreign aromatics so highly prized, in the same liquor.

Recipe 8.—Gooseberry Wine.

(Communicated for the *Archives of Useful Knowledge*, by a lady. See vol. 3, 378.)

Dissolve three pounds of white sugar, in four quarts of water, boil it a quarter of an hour, skim it well, and let it stand till it is almost cold; then take four gallons of full ripe gooseberries, bruise them in a mortar, and put them into your vessel; then pour them in the liquor; let it stand two days, stirring it every four hours, steep half an ounce of isinglass chipped fine in a quart of brandy, two days; strain the wine through a flannel bag, into a cask; then beat the isinglass and brandy in a mortar, with the whites of five eggs; whisk them together, half an hour, put it in the wine, and beat them all together; close up the cask, and put clay over the cork; let it stand six months, then bottle it off for use; put in each bottle a small lump of sugar, and two jar raisins.

This is a very rich wine, and when kept in bottles, two or three years, will drink like Champagne.

Letter from Dr. Anderson respecting Home-made Wines.

"I can say little else, than that from our own experience for a short time past, and what I have seen of others, I am perfectly satisfied that wine may be made from our native fruits—red and white currants, gooseberries, black currants, raspberries, and other fruits (with the help of sugar), as good, and of as rich a flavour in all respects, as any that are imported from abroad. But the particulars in the process that may vary the qualities of the wine, where the materials are the same, are so numerous, and the time that must elapse before the result of any experiment can be known is so great, that I despair of living to see any certainty established on this head. At present, I sometimes taste as good wine of that sort as could be desired, and again as bad as can be thought of, made by the same persons, when they can assign no reason for the difference. From our own limited practice, I have been able to ascertain only two points, that I think can be relied on as tolerably well established. These are, *first*, that age, I mean not less than *three* years is required to elapse before any wine, that is to be really *good*, can attain such excellence as to deserve the name of *good*: and *second*, that it never can attain that perfection, if spirits of any kind be mixed with it. I apprehend that most of our made wines are hurt by not adverting to these two circumstances.

"Another circumstance that is, in my opinion, very necessary for the formation of good wine of this sort, is a certain degree of *acidity* in the fruit, without which the wine never acquires that zest which constitutes its peculiar excellence, but hurries forward too rapidly into the state of vinegar.

"Currants at all times possess enough of that acidity, but if gooseberries be too ripe they are apt to want it, and become insipidly sweet at an early period, though they soon become vinegar. It ought to be remarked, that the native acidity of the fruit is different from the acidity of vinegar, and possesses qualities extremely dissimilar. The sourness of vinegar, when it has once begun to be fumed, continues to augment with age; but the native vegetable acid, when combined with saccharine matter, is gradually diminished as the fermentation proceeds, till it is totally lost in the vinous zest into which both this and the sugar are completely converted before any vinegar is produced, if the fermentation be properly conducted.

"This I believe is a new opinion, which experience alone enabled me to adopt not very long ago. But I have had so many experimental proofs of *this* fact, independent of the support it derives from reasoning, that I am satisfied it is well founded. I am satisfied farther, that the wines of this country are debased chiefly by not adverting to it, and of which I think you will be convinced also by a moderate degree of attention.

"Every person knows, that an insipid sweetness is the prevailing taste in liquors when they begin to ferment, and that it is gradually changed into a pungent vinosity as the process proceeds; but few persons have had occasion to remark, that the *native acid* of fruit undergoes a similar change by the fermentatory process. Every one who tastes made wines, however, soon after the process has commenced, perceives that the sour to a certain degree is mixed with the sweet. It chances, indeed, that the sweet is sooner blended than the sour, so that when the liquor is tasted a few months after it has been made, it hath lost some part of its sweetness, but still retains nearly the whole of the sourness of the native acid of the fruit; and as the vinous flavour is yet but weak, the liquor appears to be thin and weak, and running into acidity. It is therefore feared, that if it be not then drunk, it will soon run into the state of vinegar; on this account it is often used in this state, when it forms a very insipid beverage. Frequently also, with a view to check the acetous process, and to give that degree of strength which will entitle it to the name of a cordial liquor, a certain portion of brandy is added to it, after which it may be kept some time. The effect of this addition is to put a stop to that salutary process of fermentation which was going slowly forward, and gradually maturing the native acid vegetable into vinous liquor, which being at last blended with the saccharine vinous juice, produces that warm and exhilirating fluid which cheers the heart and invigorates the strength of man. In this way the sharp insipid and poor liquor which was first tasted, is, by a slow process, which requires a great length of time to complete it, converted into rich pleasant wine, possessing in a great degree that high zest which constitutes its principal excellence.

"My experience does not yet enable me to speak with certainty, respecting all the circumstances that may affect the flavour, or augment or diminish the strength of wine, or accelerate or retard the time of its ripening. But my opinion at present is, that a great part of the flavour of wine depends considerably upon the skin of the fruit, which may be augmented or diminished by the degree of pressure the fruit is subjected to, and other particulars connected with it; or by the macerating the fruit more or less in the juice, before the skins be separated from the pulp; and that the ultimate qualities of the wine are consideraly affected by the proportion of the original native acid of the fruit, conjoined with the saccharine part of the juice. It seems to me very evident also, that the saccharine juice can be more quickly brought into the state of wine than the acid portion of it, and that of course, those wines that consist entirely of saccharine matter, flavoured only by some pleasing vegetable perfume, such as cowslip or elder flower wine, and others of similar sorts, may be sooner brought to be fit for drinking than those in which the juices of fruit form a considerable ingredient, and may be also made of a weaker and lighter quality. And that fruit wines, in proportion to the diminution of the quantity of fruit to that of the sugar, or in proportion to the quantity of acid in the fruit, may be accelerated in the progress of fermentation; but that strong full-bodied wine, of good flavour, must have a considerable proportion of native acid, and requires to be kept a long while before it can attain its ultimate perfection."

In speaking of grape wine, Dr. Anderson mentions an experiment made with grapes that were perfectly ripe. The juice was squeezed out with the hand, as he had no press. It fermented well. The liquor, when tried, had a sweetish taste, but wanted much of the vinous zest wished for; this he supposes owing to a want of a due proportion of the native acid. Accordingly, in another experiment, he plucked the grapes rather sooner, when the juice possessed more vegetable acidity and less of the saccharine taste than when fully ripe; fearing, however, that the juice might not be sufficiently matured to do by itself, he added a portion of sugar and water to the juice. The liquor fermented well and had a promising appearance.

An useful Recipe for making Family Wine.

(*Nicholson's Journal*, 19 vol. p. 354.)

Take black currants, red do., white do., ripe cherries (black hearts are the best), raspberries, each an equal or nearly equal quantity; if the black currants be the most abundant, so much the better. To four pounds of the mixed fruit, well bruised, put one quart of clear, soft water; steep three days and nights, in open vessels, frequently stirring up the mass; then strain through a hair seive; the remaining pulp press to dryness; put both liquids together, and to each gallon of the whole, put three pounds of good, rich, moist sugar, of a bright yellowish appearance. Let the whole stand again three days and nights, frequently stirring up as before, after skimming off the top. Then turn it into casks, and let it remain full, and purging at the bung-hole about two weeks. Lastly, to every nine gallons put one quart of good brandy, and bung down. If it does not drop fine, a steeping of isinglass may be introduced and stirred into the liquid, in the proportion of about half an ounce to nine gallons.

N.B. Gooseberries, especially the largest and rich flavoured, may be used in the mixture to great advantage; but it has been found the best way to prepare them separately, by more powerful bruising or pounding, so as to form the proper consistence in the pulp; and by putting six quarts of fruit to one gallon of water, pouring on the water at twice, the smaller quantity at night, and the larger the next morning. This process finished as aforesaid, will make excellent wine unmixed; but this fluid added to the former mixture, will sometimes improve the compound.

Annotation by Mr. Nicholson.

I am inclined to think the addition of brandy here recommended, injurious; an opinion founded on the authority of a respected friend, formerly a chemist in a county town, who excelled in making family wine, and confirmed by my own experience. A similar opinion is entertained by Dr. Anderson.

I will only add, that the best home made wine I recollect to have tasted, was made by expressing the juice of white currants, bruised but not picked from the stalks, adding water to the fruit after it was pressed, in the proportion of double the quantity of juice; mixing the two liquors together, and putting the whole into a barrel with three pounds of pretty coarse, brown sugar to every gallon of the mixture, stirring it well, and leaving it to ferment with the bung-hole, at first open, and afterwards loosely covered, the barrel not being quite filled. As the sugar does not immediately dissolve, the stirring must be repeated at intervals of a few days, till this is effected. After it has fermented properly, the barrel must be stopped close; and it may be afterwards bottled for use.

Chapter XI.

To make Beer.

Receipt to brew one bushel of Malt.

First procure a large tub with a false bottom, in which are bored a number of holes that will not let the malt through, put into this tub one bushel of malt coarsely ground, then add ten and a half gallons water (say three of cold water, and seven and a half of boiling water) stir it well and let it stand three hours, draw it off by a hole through the bottom; then add six gallons more of water to the malt, a little hotter than the first, stir it well and let it stand two hours, draw it off; then put on four gallons more of water, hotter than the last, let it stand one hour, then draw it off.

The two first liquors should now be put on the fire to boil, with half a pound of hops, and continue to boil two and a half hours, as it boils down to be filled with the last wort, so as to make in the whole twelve gallons, which must be strained through a hair seive, and set in tubs to cool; when it gets to seventy degrees of heat, add to the twelve gallons, a tea cup full of sweet yeast (brewer's yeast if possible), then put it into a keg, and stand the keg on a tub to save the beer that works out of the bung hole, fill up the keg three times through the day, and in two days and a half, it may be bunged up and put by, and in ten days it will be fit for use.

Chapter XII.

Concerning Cider.

The process of making cider is so simple, so generally practised, and considered to be so well understood in every part of the United States, that any observations on the subject would seem to be almost superfluous. Unfortunately however, this very simplicity and general knowledge of the operation, tends to ruin three-fourths of the cider that is made.

The increase of orchards, and the real value of cider itself, have rendered it an article of considerable importance, not only as an object of commerce, but as a valuable beverage for home consumption. To the farmer himself, a mug of sweet cider is frequently considered as a great luxury, and by labourers it is preferred by way of breakfast, to tea, coffee, or milk; and in the harvest field to the more intoxicating liquors generally prepared for them.

Yet notwithstanding this great utility and general use of cider, how seldom do we find it amongst the majority of farmers, fit to drink in the month of February. To what cause can this be attributed, when on the other hand we sometimes find, cider that has retained its original flavour and sweetness for 18 months?

An exposition of the causes, the method of avoiding them, and some directions for keeping cider, we trust will not be deemed improper, and may possibly lead to some improvement in this subject so important to many farmers.

In the first place, the mill is not perfectly cleaned, previously to grinding the apples; next the apples are picked green, ripe, and rotten, as it may happen, and together with a quantity of grass and leaves, are all ground up; and lastly, the liquor without straining (and consequently with a good deal of pomage in it), is put into dirty casks, the bung being loosely stopped with straw, and rolled away into the cellar, where no further attention is thought necessary until the cider is wanted for use; when, in consequence of this very improper mode of treatment, it is found perfectly sour.

English writers on this subject, give a great many directions, but as they all require more labour than we can command, and are of little use in this country; the time for making keeping cider, occurring in one of the busiest seasons of the year.

What however may be usefully mentioned, and agrees with our own experience, are the following general directions:

That the apples should be as nearly as possible of an equal degree of ripeness, and if not perfectly ripe when gathered, should be put in a heap for a few days to mellow,* when ground, care should be taken that every thing be perfectly clean, and the straw used in making the cheese, should be free from must, or any disagreeable smell which might be imparted to the cider.

The pomage should remain from twelve to twenty-four hours after grinding, before it is pressed; the cider must be carefully strained and put into clean casks, avoiding new ones, unless made of perfectly seasoned wood, or such as have had any liquid in them which might flavour the cider. Here may be found the grand stumbling block of most farmers.

When a cask of cider is run out, there will generally be left a few gallons of lees: the bung and spicket hole are left open, and in this situation it frequently remains until the next cider season, when, after a few scaldings, which are of little effect, it is filled with fresh cider; the inevitable consequence is, the cider will become sour. To avoid this, so soon as a cask is out it should be completely emptied, and scalded perfectly clean, or well washed with lime water, dried and bunged up close; it will be then found sweet when wanted.

Much has been said in favour of racking off cider frequently, to prevent acetous fermentation, and though I am unwilling to differ from the high authorities upon which it is recommended, as the result of various experiments, yet I cannot but offer a few remarks on the subject.

When cider is made late in the season, so as to undergo a slight fermentation, sufficient however to give it an agreeable vinosity, it may become a question, whether by suffering it to remain on the lees, they do not afford a kind of *feed*, by which it retains its strength and vinosity longer than it otherwise would do?

This however is more particularly applicable to cider for immediate use, it certainly being proper to rack off from the lees, that which is meant for keeping, some time in the month of March, when the weather is too cold for fermentation, to which it is liable upon being agitated.

It is also worthy of attention whether cider, racked off upon the subsiding of the first fermentation, which has been slight, does not immediately undergo another fermentation (in consequence of the agitation and mixture of the fermentable principle which subsides with the lees), which although almost imperceptible, in a short time renders the cider sour.

Especial care must be taken to fill the barrel to the very top of the bung hole, at the last racking; that if any light or flying lees remain in the liquor, they may be removed at the bung, for this is frequently the case of mellow cider; and if those lees are permitted to remain in it, the surface, by being exposed to the air, will become sour. That tartness will by degrees render all the cider of the same complexion. Yet, the taint may be perceived to descend gradually; for while the cider is sour at the top, it is sound a few inches below it, till it descends from top to bottom. This is the grand article in which people are wont to be deceived, and by which they are rendered out of humour,

*Apples should be housed to keep them from rain; and the whole process of cider making carried on under cover.

with racking of cider, how much soever they are pleased with it, when it happens to answer their wishes. When their cider turns sour, they imagine that racking takes away the spirit of it, and that it must then become sour of course, for want of a body, as they are wont to speak; whereas in truth, it grows sour for want of skill how to secure it after the last racking, by removing the light lees which swim on the top, before they acquire the last degree of acidity.

The commencement of acidity may be known by a singing or hissing noise; this should be immediately attended to, and probably the most effectual and certain remedy, will be in the addition of a small portion of high proof spirit, and the bunging the cask tight.

Cider put into the cellar so soon as made, generally undergoes too great a fermentation. To prevent this, when it is made late in the season, the cask should be placed in the north side of a house and completely protected from the sun; the warmth of the day disposes the cider to ferment, but the coldness of the night so far checks this disposition, that only a slight yet a complete vinous fermentation takes place.

When however this is completely stopped by the increased coldness of the weather, and before it freezes, the cider should be run off into casks placed in the cellar, with as little agitation as possible, and about one gallon of brandy added to each hogshead; the casks then being closely bunged, no further fermentation will take place. This change of temperature, is a powerful opponent to fermentation. But strong sweet cider, put into a cellar where there is a constant uniformity of temperature, even though it be very cool will almost certainly ferment, and the fermentative principle once completely in action, can scarcely be stopped, but by a very great increase of cold.

I am well aware, that the ideas here advanced as to the necessity of cleanliness, in every part of the operation, but particularly in the casks, will be *ridiculed* by many; having experienced the fact that very fine cider has been produced where no further pains were taken by the owner, than to order an old negro with two or three boys, "to make the cider," and the casks were probably, only washed with a little cold water. This however might only happen once in half a dozen years, and should be regarded as an accidental concurrence of circumstances, probably beyond the art of man to elucidate. It is an exception to a general rule, and in direct opposition to theory and correct principles; such accidental circumstances, therefore, should not be regarded by the man who wishes to act according to system.

When cider is wanted for making wine or any particular use, the last running from a pressing should be taken, as this will be found more pure and perfectly free from pomage.

Receipt to make Cider.

(Agreeably to the plan practised in Ireland.)

After the apples are bruised and pressed in the usual manner, the juice should be immediately put into large open vessels, and suffered to remain in this situation from twenty-four to forty-eight hours, in order to deposit any crude matter which may have passed through the bag; and also to throw up the lighter particles in the form of scum, which should be carefully removed; the liquor is then to be drawn off and passed through a double flannel bag, removing the feculent matter by occasionally turning and rinsing it. When thus prepared, put two or three gallons into a strong well bound cask, in which matches (made by dipping linen rags in melted sulphur), are to be lighted and suspended from the bung hole, by means of iron wire, and the bung lightly put in, fresh portions of match must be added until they cease to burn on their being introduced into the cask, which should now be violently agitated for the purpose of assisting the absorption of sulphurous gas. After standing a quarter of an hour, draw it off into a tub, the cock and bung being left open, that the light unabsorbed gas may be suffered to escape; after remaining in this situation for about fifteen or twenty minutes, the operation must be repeated five or six times, with a like quantity of fresh liquor each time; return the different portions into the cask, and fill it up with filtered liquor; put a quart of spirits to every forty gallons, and insert the bung in the firmest and closest manner, so as to preclude the possibility of the internal air forcing a passage, should it be disposed to ferment. In six months it will be fit for bottling, the corks must be wired down and the bottles laid on the side in binns.

Rationale.

Sulphurous acid, which is formed by burning sulphur in confined portions of atmospheric air, has the well known property of checking fermentation, so that if the fresh juices of fruit be impregnated with this acid, it causes a suspension of the vinous fermentation, until sufficient time is afforded for the forming of the liquor, which on its being bottled, gradually ferments, and causes it to assume that fine sparkling appearance met with in English cider. In the common sour cider, the fermentation has proceeded through the vinous to the acetous, and consequently in a state nearly approaching to vinegar.

In the mode usually practised, in making cider, in imitation of English, the fresh juice is at first put into the cask, and the whole drawn off when it shews signs of fermentation, the casks stoved with sulphur match, and the liquor immediately returned; racking off, and stoving it until it ceases to exhibit a disposition to ferment, which tedious process usually takes about six weeks, but which may be advantageously shortened by the substitution of sulphur matches, in larger proportions as before directed; so as to impregnate the liquor with sulphurous gas.

Chapter XIII.

Improvements, or Substitutes, for the common Worm.

Mr. Acton, of Ipswich, in England, having used a still, containing nine gallons, for distilling common water, essential oils, and water, refrigerated them into a tub, containing about thirty-six gallons, found it very inconvenient to change the water of the tub, as often as it became hot, which it very soon did, after commencing distillation; he therefore contrived the following addition to the refrigerating part of the apparatus, which he has found to succeed so well, that he can now distil for any length of time, without heating the water in the worm tub, above one degree; so that it never requires to be changed; the heat passes off, entirely into the additional condenser, and when it exceeds 150° goes off by evaporation. The additional condenser consists of a trough, three feet long, twelve inches deep, and fifteen inches wide, with a pewter pipe passing through the middle of it horizontally, about two inches in diameter at the largest end, next the still, and gradually tapering off to about three quarters of an inch at the smallest end, which communicates with the worm. The great simplicity of this contrivance, and its utility, render a fair trial of it in other stills advisable; the small degree of heat, which went to the water in the worm tub shews, that the additional condenser performed nearly the whole of the condensation, and that therefore it is extremely probable, that a second pipe and trough added to the first, would perform the whole condensation effectually, without using any worm, and thus enable distillers to dispense with this expensive and troublesome part of the apparatus.

Remarks by H. H.

Had Mr. Acton lengthened the worm three feet, increased the size of the worm tub, and added a quantity of water equal to the contents of the trough he used, I think he would have been equally successful. Still, however, it is a matter worthy of attention, and I trust we shall have some satisfactory experiments on the subject before long, as it is a favourite project with several gentlemen of this country. I never thought that the same quantity of water applied in a horizontal trough could be equal to the old plan of a worm, but it would probably be of advantage, to have a horizontal semicircular tube in place of a worm for the waste still, on account of the ease with which it might be cleaned, in case of being choked. The difference of expense is scarcely worthy of consideration; but by placing the troughs out of doors, the room of the worm tub would be gained.

A worm made of semicircular pipes, with the flat side below would cause a greater surface to be exposed to condensation.

Another substitute for a Worm.

Procure two of the largest and widest sheets of copper. Let two frames of wood be constructed of such dimensions, that the sheets may be turned over the edges and nailed. Stretch the copper by a mallet, as evenly as possible, so that it may *bulge* towards the opposite side from where it may be fastened. Clamp the two frames together by screws. By means of two holes at opposite corners, the steam may enter and pass out when condensed.

Another substitute.

M. Lapadius of Frieberg, has discovered a method of condensing vapours in distillation, more rapidly than has yet been done. This is accomplished by means of a disk, attached to the tube of the still, which has the figure of a lens, flattened as much as possible, and is made of copper. It produces a much better effect than the worm hitherto employed for that purpose.

Description of Baron de Gedda's Condenser.

(See *Rep. of Arts*, vol. 21, New Series.)

This condenser consists of two cones, obtruncated and reversed, placed one within the other, leaving between them an interval closed at top and bottom by rings soldered to the cones. It is in this space, which is three times larger above than below, that the condensation of the alcoholic vapour is effected. The interior cone being truncated, lets the water of the refrigerator pass, which striking the interior and exterior surfaces of the conical condenser, speedily cools the liquor. The upper diameter of the exterior cone is to its lower diameter as seven to four. The height of the cones is to their greatest diameter, nearly as five is to two. The least diameter of the interior cone is to that of the exterior cone as 18 is to 21, and the difference of their greatest diameters as 21 to 30. So that in the largest condensers, which are about six feet high, and are used for alembics containing 100 cubic feet, the interval between the cones at the bottom is only an inch and a half, while the space between them at the top is about five inches. The condensers of the least dimensions are formed with the same proportions.

It will readily be perceived, that a tube passes from the upper part of the condenser through the keeve, to form the connection with the still; and that another tube proceeds through the keeve likewise, from its lower extremity, to transmit the liquor.

The advantages of this condenser are, that from its particular form a complete and rapid condensation takes place in the upper part of it, and the liquor runs off below extremely cold; it is more easily made, takes less materials, and is consequently less expensive than a worm of the common sort, it is more durable, not liable to the same accidents, and more easily kept clean; since it should be so contrived that the top may be taken off, when it can be cleaned with a brush through its whole extent.

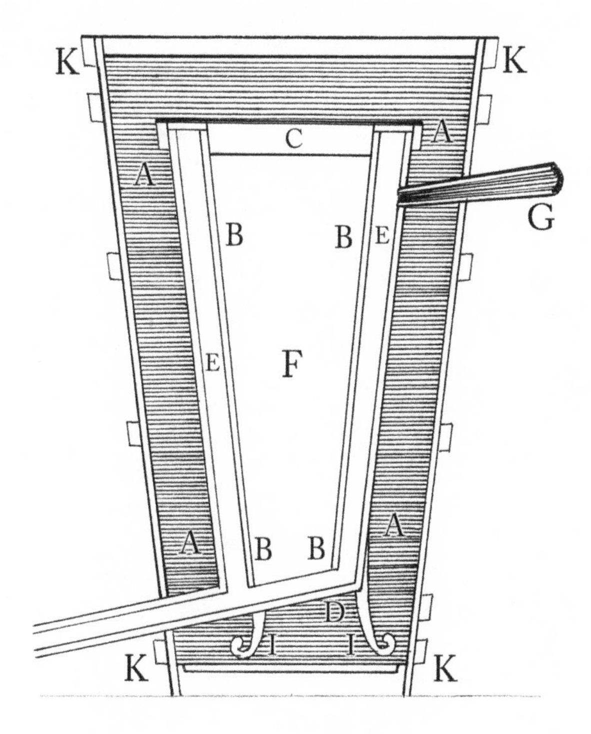

Explanation—The figure represents a vertical section of the conical condenser, and of the vessel in which it is placed.

A A A A the exterior cone.

B B B B the interior cone, made of copper or tin.

C the ring that closes the interval above.

D the ring that closes the bottom of the interval, these rings are soldered, and serve to unite the two cones.

E E the space between the cones where the condensation of the vapours takes place.

F the open space at the lower part of the internal cone through which the water of the refrigerator passes to cool the internal part of the condenser.

G a tube through which the spirituous vapours pass from the still to the condenser.

I the feet of the condenser three in number.

K K K K the great keeve or refrigerator filled with water.

Chapter XIV.

On Raising Water.

Any machine for the purpose of saving labour, which is not too complicated in its structure, nor expensive in its first cost, is of importance to the citizens of this country in general. One for raising water it particularly so to the distiller.

The screw of Archimedes, and the hydraulic ram, have long been celebrated in philosophy. The theory upon which each is founded is correct, and no doubt it may be reduced to practice; whether upon a scale sufficiently large to be useful, may be a subject of experiment.

The siphon is a well known instrument for drawing liquors. It may also be used to supply a flake stand in a distillery, provided there be a fall of one or two inches from the place the water is taken up to where it is discharged, that it be not raised more than 34 feet, and the flake stand be made air tight, and kept always full.

A patent has been obtained for something on this principle, which I believe succeeded very well.

The following descriptions, with the plates accompanying, will enable any ingenious mechanic to try the experiment.

A B (Fig. 1, p. 192), is a box made of thin planks, which, together with its two tubular arms B C, and B D, is moveable about the centre G and H. At the extremity of each arm is a hole, as E, of such a size that the quantity of water discharged by both, shall be less than that which falls through M N. The tube A B, will therefore be constantly full. H is the axis of the whole, fastened to the interior of A B. At I is a perpetual joint, with the upper axis of which is connected the cylinder S T, with a tube coiled round it, so as to form a screw of Archimedes. The cylinder is prevented from slipping downwards by a shoulder on its axis at a, and is moveable on the centres *a* and *b*.

The action of the machine, which is easily understood, is as follows:—the water flows through R P, which is connected with the stream by means of a pipe into the box X, until it fills it as high as K. It then continues its course through the tube K L M N, and falls into the box A B. As the holes E and F, discharge less than the quantity of water which flows into A B, it will soon become full, when the water rushing out at the two holes will cause, by its reaction, the arms, and of course the box A B, to move in a

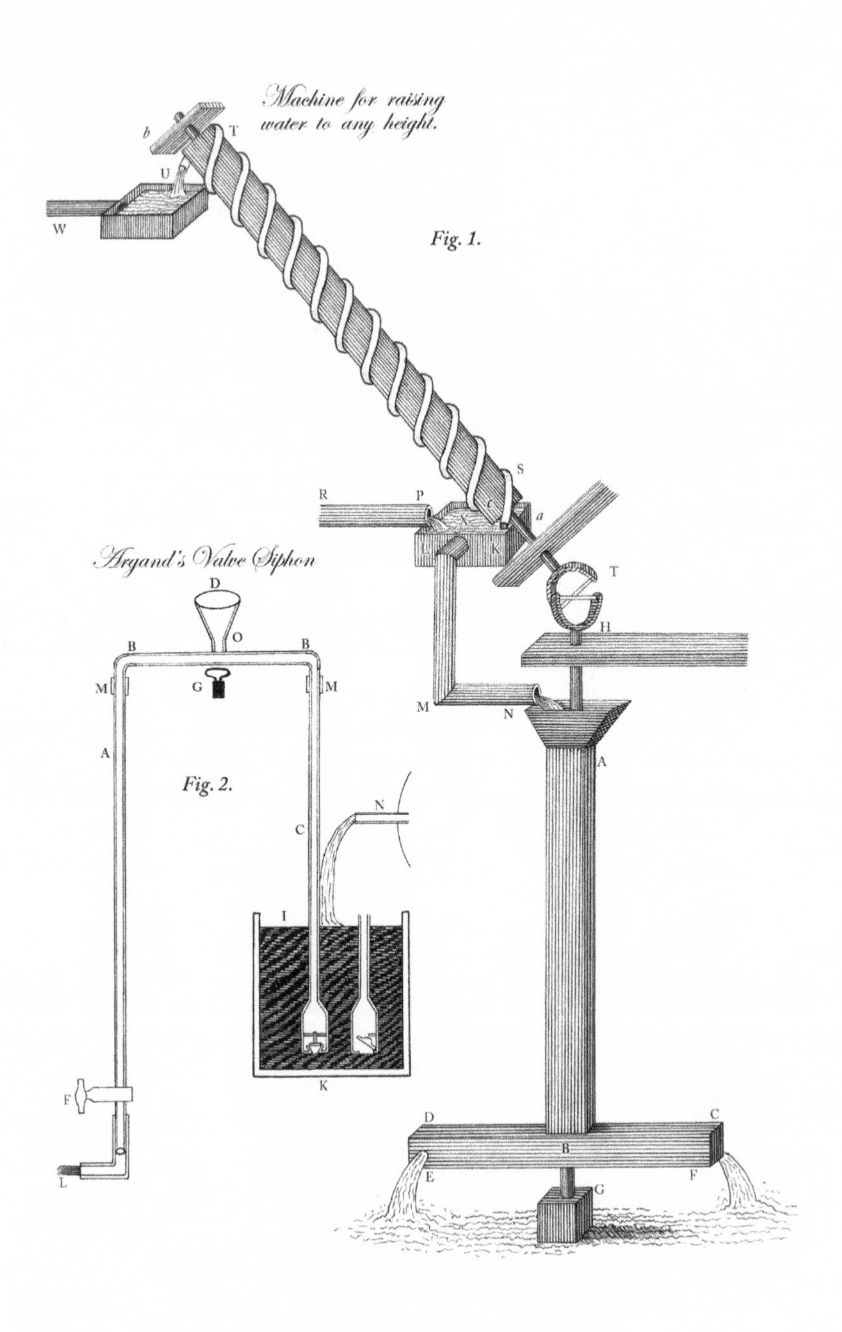

Emporium of Arts & Sciences, Vol. 2. pl. 1.

retrograde manner. The axis H revolving also, turns the cylinder S T, at every revolution of which the orifice of the pipe e descends into the water x, takes in a small portion, and as the whole turns, raises it gradually to the top, where it is discharged at U, and carried off by the pipe W.

The cylinder may be prolonged to any height desired, but it is evident that the longer it is, the smaller must be the diameter of the tube, in order that the same force may move it in both cases.

Description of the valve Siphon, of the late Mr. Ami Argand, inventor of the lamps with a double current of air.

This improvement, though simple, is ingenious, and particularly adapted to large siphons, that require to be removed from one vessel to another. A valve as E, or H (Fig. 2, p. 192), is applied to the foot of the shorter or ascending leg of a siphon A B, B C; at the other foot of which a stop-cock F, is placed. The cock being open, and the foot E immersed in any liquid in a vessel I K, by moving the leg E perpendicularly downward and upward, the liquid will gradually ascend through the valve E, till it runs out at the point L. The pressure of the air on the surface I, will then be sufficient, to force the liquid through the valve E, as long as this remains beneath it; and thus it will continue to act as a common siphon, and the vessel will be emptied, unless supplied from some reservoir as N.

As soon as the siphon is filled and begins to discharge the liquid at L, or at any period while it continues full, if the cock F be turned so as to stop it, it may be very safely and conveniently removed to any other vessel, as the cock will prevent the liquid from running out at one end, and the valve at the other; and the moment the extremity E is immersed in the liquid in another vessel, and the stop cock K turned, it will act again as before.

The siphon may be filled in this way in a clear liquid, and then removed into a vessel of the same kind of liquid, that has a sediment at bottom, which would be disturbed by moving it up and down. This however, may not always be convenient. Mr. Argand therefore, makes an aperture with a short perpendicular tube O, in the horizontal branch B B, through which, by means of a funnel D, the siphon may be filled, while the cock F is shut, so that it may be inserted into the liquid, and made to act without disturbing it. When the siphon is thus filled, or when the funnel D is not required, the aperture at O is closed by the stopple G.

For the convenience of carrying the siphon, as well as for packing it up, or cleaning it, the horizontal and perpendicular branches are made to take asunder at the joints M M. The nozzle L is likewise made to take off, as it is frequently more convenient for the fluid to be drawn off perpendicularly.

Description of the Hydraulic Ram, by Mr. Millington

The *Belier Hydraulique*, or water ram, as this machine was called by Mongolfier, who first constructed it about 1797, is applicable to any situation in which there is a fall of a few feet of clear water, and drainage to get rid of the superfluous quantity; and

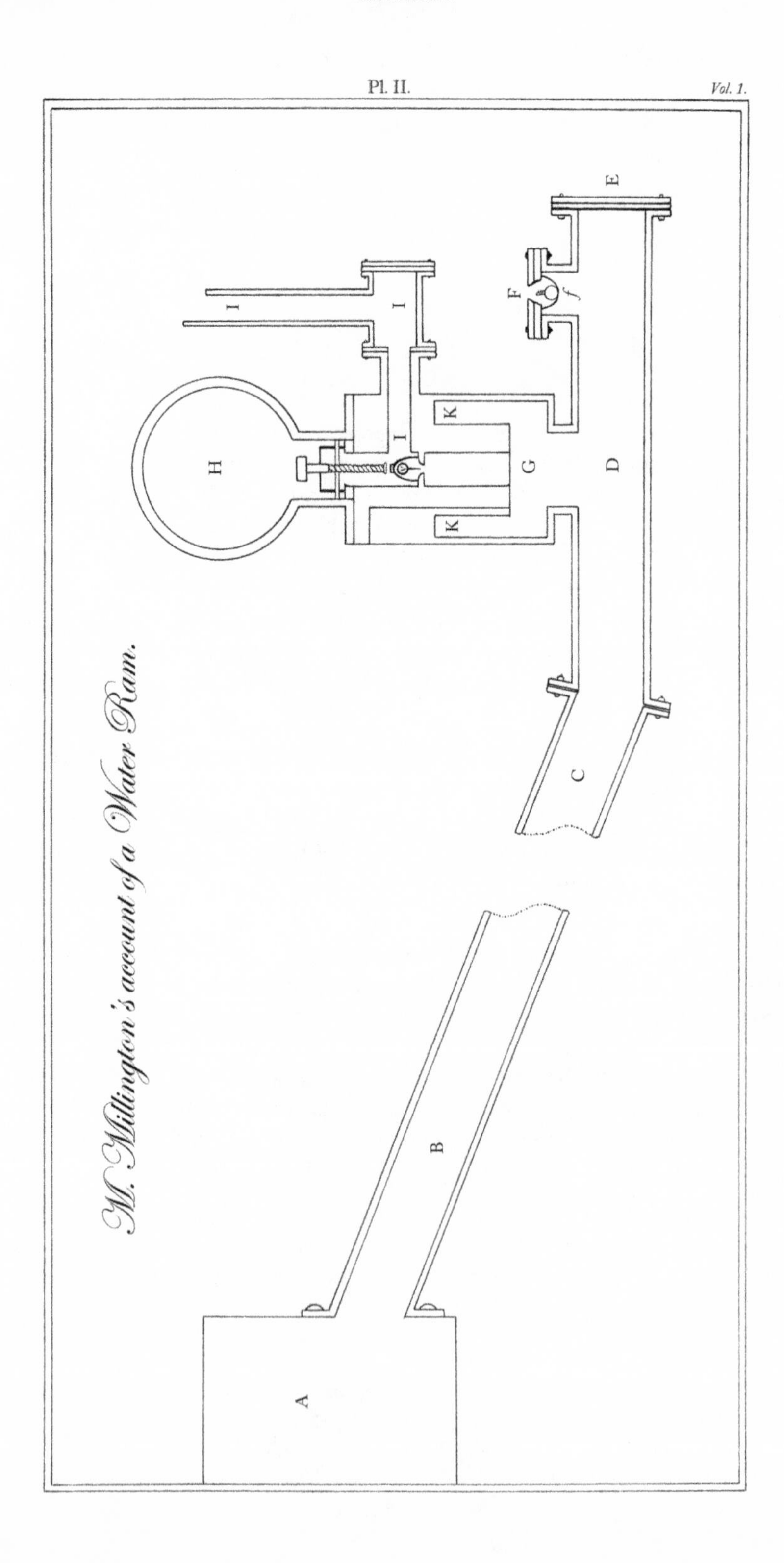
Pl. II.
Vol. 1.
M. Millington's account of a Water Ram.
A
B
C
D
E
F
f
G
H
I
I
I
K
K

as it is simple and cheap in its construction, and requires no attendance after it is once adjusted and set to work, it is particularly applicable to the supply of houses or gardens, and pleasure grounds situated upon elevations.

The action of the water ram, as will be seen in the following description of it, is entirely dependent upon the momentum which water, in common with all other matter, acquires by moving, a circumstance which has often proved very detrimental and troublesome to plumbers and others, in fixing pipes connected with elevated cisterns.

It may have been observed by many, on turning a cock attached to a pipe so circumstanced, that the water flows with great violence; and upon shutting it off suddenly, a concussion is felt, the pipe is shaken, with a noise resembling the fall of a piece of metal within it, and the pipe is not unfrequently burst open near its end.

This arises from the new energy the water has acquired, by being put in motion and then stopped, in consequence of which it makes a considerable mechanical effort against that end of the pipe which opposes its further progress.

This effect was experienced in a great degree at an hospital in Bristol, where a plumber was employed in fixing a leaden pipe, to convey water from the middle of the building to the kitchen below, and it was found, that nearly every time the cock was made use of, the pipe was burst at its lowest end; after making many attempts to remedy this evil, it was at last determined to solder a small pipe immediately behind the cock, which of course was carried to the same perpendicular height as the supplying cistern, to prevent the water running to waste, and now it was found, that on shutting the cock, the pipe did not burst as before, but a jet of considerable height was forced from the upper end of this new pipe. It therefore became necessary to increase the height of the pipe, to overcome, if possible, this jet, and it was carried to the top of the building, or twice the height of the supplying cistern, where, to the great surprise of those who constructed the work, the jet still made its appearance, though not in such considerable quantities, and a cistern was placed at the top of the house to receive this superfluous water, which was found very convenient, particularly as it was raised without trouble or exertion. This is, I believe, the first water ram which ever had existence, the circumstance having taken place prior to Mongolfier's contrivance, though he is the first person who organized, the machine and made it completely self-acting, without ever turning a cock. His construction is represented in the plate, where A is a cistern (or part of a running brook, which may be dammed up to make a head of water), and B, C, a quantity of iron or wooden pipes, extending from 18 to 30 or 40 feet in length, according to their diameter, to conduct the water away; these pipes are laid in a sloping direction, so as to reach the greatest depth D, at which the water can run off, which may be from one to six or eight feet below the head A. The water would naturally run to waste from the end E of these pipes, but that is closed by a blank or solid flanch, and it is only permitted to escape through a solid hole in the centre of the horizontal flanch F, from whence h will run in an uninterrupted stream. This hole is, however, equipped with a valve within it, as at F, and this valve is so adjusted as to sink by its own weight in the water, while that water is motionless or moving slowly.

Now if we suppose the pipe B, C, D, to be supplied with water from A, that water will at first pass round the valve, and discharges itself at F; but as soon as it has acquired a small additional force by moving, H will be more than equivalent to the weight of the valve F, and will lift it, by which the passage of the water becomes instantly stopped, and an effort will be made to burst the pipe D; this is prevented by the second orifice over the letter D, communicating with the chamber G and air vessel H, from whence there is an immediate communication by the pipe III, with the elevated situations to which the water is to be thrown. As the effect of the blow which the water makes is instantaneous, it becomes necessary to place a second valve between the air vessel and the chamber G, but below the pipe II so that any water which is thrown into H by the effort, may be confined there, and acted upon by the condensed air, instead of permitting it to return and equalize itself in the pipes C, D. The blow which the water makes is so sudden and violent as to produce an expansion in the pipe D, which is as suddenly succeeded by a recontraction and trifling vacuum in D, by the tendency of the water to return up to C when stopped; the effect of this is to bring down the valve F, by which a free passage is once more opened for the water, which again flows and shuts F as before, to produce another blow or pulsation, by which a second quantity of water is thrown up II. Each repetition of this operation affords a fresh supply of water.

It will be evident that the valve F, as well as V, will require some adjustment as to weight. This is effected by making these valves of hollow brass balls, having a hole on one side, by which some shot or small piece of metal can be introduced to adjust the weight. The hole is afterwards stopped by a screw, which projects and forms a shank or tail to guide the valve. The screw over V is likewise to adjust the height to which that valve should rise, and to prevent its breaking away and getting into the air vessel, which it otherwise might do from the violence of the blow.

It has been found, that after using the water ram for a short time, as it was formerly constructed, the air in H became absorbed and entirely disappeared* and by its ceasing to act as an air vessel, the water would not proceed to any great height up 11. This is obviated in the present case by the chamber O placed between the air vessel and the pipe D. From the form of this chamber, any air which enters it becomes confined in the recesses K K, and not only equalises the action on the valve V, but makes the whole motion less instantaneous. K K becomes supplied with air in small machines, by the falling of the valve F, which brings a small quantity of air down with it. In larger ones, it will be necessary to apply a small shifting valve, or spring valve opening inwards to some part of the outside of G, when the air, as it enters, will rise to the top of K K, and as it accumulates, will at length pass through A into H, and keep it supplied with air.

This latter contrivance, I believe, originated with Mr. Dobson, of Mortimer-street, Cavendish-Square, who has paid considerable attention to the improvement of this engine, and proposes erecting them for the public. In the rams which I have seen, the tubes B, C, D, have becu from 1½ inch to 4 inches diameter, and the ascending pipe II, one inch, or rather less. I have seen the valve F make from 50 to 70 pulsations in a minute, and I should think discharging near half a pint of water at each pulsation, at the height of 30 feet, with a six feet head. I am, however, told, that a machine has been made which furnishes one-hundred hogsheads of water in 24 hours, to the height of 134 feet perpendicular, with a fall of four feet and an half. I am not aware that the best proportion of parts has yet been ascertained, or the quantity of loss compared with the quantity delivered up II, which must in a great measure depend upon the heights of the respective heads, and the size and length of B, C, compared with the perpendicular fall from A to D. I intend entering into an examination of these points, and if you should think the result of my inquiries worth inserting in a future number of your journal, they shall be very much at your service.

Chapter XV.

Of the Thermometer.

The thermometer was invented by Sanctorius, an Italian physician, about the beginning of the 17th century; but it was of little use until improved by Mr. Boyle and Sir Isaac Newton.

Thermometers are made putting mercury into small glass tubes with bulbs, and heating these bulbs until the mercury boils. This ebullition exhausts the tubes of air, and they are hermetically sealed while the mercury is boiling; which preserves the vacuum. The bulb is then immersed in ice or snow, and the point to which the mercury falls is called the freezing point. The intermediate distance is afterwards correctly graduated.

Thermometers filled with alcohol are useful for ascertaining very low temperatures, in which mercury would be frozen; it is only by the most intense cold that can be produced that alcohol can be frozen. For very delicate experiments *air* thermometers are used, in which as the air is expanded or contracted, a coloured liquor is made to rise or fall, which marks the degrees of expansion, and consequently the variation of temperature, they are called thermoscopes.

Fahrenheit's thermometer is universally used in this country and in Great Britain. In it the range between the freezing and boiling points of water is 180°, and as the greatest possible degree of cold was supposed to be that produced by mixing snow with muriate of soda; that was made the zero; thus the freezing point became 32°, and the boiling point 212°.

The centigrade thermometer of modern France places the zero at the freezing point, and divides the range between it and the boiling point into 100°.—This has long been used in Sweden under the name of Celsius's thermometer.

Reaumer's thermometer which was formerly used in France, divides the space between the freezing and boiling point into 80°, and places the zero like the centigrade at the freezing point.

De Lisle's thermometer is used in Russia. The graduation begins at the boiling point, and increases towards the freezing point; the boiling point is marked at 0, and the freezing point at 150°.

To measure the degrees of heat in high temperatures Mr. Wedgwood contrived a very useful instrument, which he called a pyrometer, including a range of nearly 32,000 degrees of Fahrenheit, a description of which may be found in the 72d vol. of the *Philosophical Transactions of the Royal Society of London*. With the death of Mr. Wedgwood the secret for making the species of clay for these pyrometers has been lost.

In Wedgword's pyrometer, the zero corresponds with 1,077° of Fahrenheit, each degree of which is equal to 130 of Fahrenheit. Hence we have the following comparison: 180° F = 100° C = 80° R = 150 D. L.= 18/13 W.

The thermometer is a useful, indeed necessary part of Dicas's or Atkins's hydrometer, as will be seen by a reference to the description of that instrument.

In the brewery and distillery it ought always to be in the hands of the manager as no process can be correctly attended to without, and it forms the only certain mode of ascertaining the heat of various mixtures.

Chapter XVI.

Of the Hydrometer.

THE hydrometer is an instrument for the purpose of ascertaining the specific gravity of fluids. It is used by distillers in reducing their spirits to proof, and by brewers to point out the richness of their extract. In philosophical experiments, where very great nicety and precision are required, it may be somewhat inferior to the hydrostatic balance, but for every purpose of the distillery and brewery, and for the general purposes of commerce, it is sufficiently accurate. On account of its simplicity, and the facility of using it, we recommend it to our readers.

Hydrometers of various constructions are in use in different parts of Europe, and probably the best for common purposes is the one invented by Mr. I. Dicas, of Liverpool. This has been adopted as the standard by the United States, and by the state of Pennsylvania. These instruments are made by Mr. Fisher, North Second-street, Philadelphia. And as a particular description, and instructions for their use, accompany each, it is not necessary in this place to insert either.

In New-York, Southworth's hydrometer is the standard. It is more simple in its construction than Dicas's, though liable to the objection of being calculated for fluids only at a temperature of 60° Fahrenheit. Hence the liquid must be at that temperature, or some additional calculations will be requisite. Those who are in the habit of using this hydrometer state it to be equally accurate, and that in careful experiment it will agree very nearly with Dicas's. Directions for using this instrument accompany it.

We have more than once, in the preceding pages, mentioned the *necessity* of an hydrometer to every distiller. To those who use it in reducing alcohol we will merely observe that on mixing a portion of alcohol and water together, the mixture becomes turbid, as if a precipitation had taken place, after a few seconds it will be seen that this is owing to the disengagement of a quantity of gas, which had previously been held in solution by one of the ingredients, which rising through the liquor, clears it from the bottom. On account of the disengagement of this gas, which is specifically much lighter than the alcohol or water, and because of their intimate union, the mixture will be specifically heavier than the mean density of their previously ascertained specific gravities. This is what has been termed the concentration of liquors, and takes place to so great a degree that 18 gallons of water and 18 gallons of alcohol on being mixed, will only measure 35 gallons. Therefore the loss on reducing spirits of such a strength to proof will be about 3 percent, unless attention be paid to this fact—for it is evident that the strength of any proposed mixture of alcohol and water cannot be calculated *a priori*, but is only to be known by actual experiment.

This will be made more plain by the following extract from a very valuable set of tables, published in the year 1794, by Mr. Gilpin in the *Philosophical Transactions*.

Extract from Mr. Gilpin's Table; Phil. Trans. 1794.

Water.	Alcohol.	at 30°	40°	50°	60°	70°	80°
10	0	1.00774	1.00094	1.00068	1.00000	.99894	.99759
10	1	.98804	.98795	.98745	.98654	.98527	.98367
10	2	.98108	.98033	.97920	.97771	.97596	.97385
10	3	.97635	.97472	.97284	.97074	.96836	.96568
10	4	.97200	.96967	.96708	.96437	.96143	.95826
10	5	.96719	.96434	.96126	.95804	.95469	.95111
10	6	.96209	.95879	.95534	.95181	.94814	.94431
10	7	.95681	.95328	.94958	.94579	.94193	.93785
10	8	.95173	.94802	.94414	.94018	.93616	.93201
10	9	.94675	.94295	.93897	.93493	.93076	.92646
10	10	.94222	.93827	.93419	.93002	.92580	.92142
9	10	.93741	.93341	.92919	.92499	.92069	.91622
8	10	.93191	.92783	.92358	.91933	.91493	.91046
7	10	.92563	.92151	.92723	.91287	.90847	.90385
6	10	.91847	.91428	.90997	.90549	.90104	.89639
5	10	.91023	.90596	.90160	.89707	.89252	.88781
4	10	.90054	.89617	.89174	.88720	.88254	.87776
3	10	.88921	.88481	.88030	.87569	.87105	.86622
2	10	.87585	.87134	.86676	.86208	.85736	.85248
1	10	.85957	.85507	.85042	.84668	.84092	.83603
0	10	.83896	.83445	.82977	.82500	.82023	.81530

Chapter XVII.

On Russian Distillation.

An interesting treatise was published at Moscow in 1792, by Mr. Vassili Nicolævitch Subof, the director of economical affairs in the government of Pensa. This gentleman, who possesses large distilleries in the district of Saratof, and has paid much attention to the subject, describes the means of increasing the quantity of spirits in distillation. He not only offers, as he had previously done in the newspapers, to *admit pupils into his school* for improving the art of distilling, but he likewise informs the public of the principles on which these improvements essentially depend. Thus he has, with the common apparatus, from a tchetvert of corn weighing nine pood, or 360 Russian pounds, produced six eimers and a quarter of spirits;* while others, from the same quantity of grain, could obtain only five eimers. This remarkable increase of the spirituous produce, the author principally attributes to the following circumstance: In order to reduce the temperature of the hot water used in the mash, he caused cold water and ice to be added, by which the evaporation of spirituous parts during fermentation was prevented. By this and some other practical advantages, which he promises to communicate without reserve to his pupils, he has brought this art to such perfection, that from 10 pood or 400 Russian pounds, he uniformly obtains seven eimers and four-fifths of common proof spirits.

An eimer is thirteen quarts English wine measure.

*About four gallons to the bushel.

Chapter XVIII

On the preservation of Vegetables for distillation by salting.

To preserve rose leaves; take 4 pounds of rose leaves and pound them two or three minutes with one-third of their weight of common salt. The flowers bruised with the salt will soon give out their juice and produce a paste of little bulk, which must be put into an earthen vessel, or small cask, and proceed in the same manner until you have filled it. Then stop the vessel close and keep it in a cool place until wanted. This fragrant paste you may distil at leisure, in a common still, diluting it with about double its weight of pure water. This process is particularly applicable to those herbs, the water of which distilled in the common way will not keep.

Chapter XIX.

Ottar of Roses.

This fragrant perfume which is sold in the East Indies at the exorbitant price of *twenty guineas per ounce*, may be made as follows:

Let a quantity of fresh roses be put into a still, with their flower cups entire, together with one-third of their weight of pure water. The mass is now to be mixed with the hand, and a gentle fire kindled beneath. When the water becomes hot, all the interstices must be well luted, and cold water placed on the refrigeratory at the top. As soon as the distilled water comes over, the heat should be gradually diminished, till a sufficient quantity of the *first runnings* be drawn off. Fresh water is then to be added, which should be equal in weight to the flowers, when the latter were first submitted to the still; and the same process repeated, till a due proportion of *second runnings* be procured. The distilled water must next be poured into shallow earthen, or tin vessels, and exposed to the air until the succeeding morning, when the *ottar* or *essence* will appear congealed on the surface. The latter is now to be carefully skimmed, poured into phials, and the water strained from the lees, should be employed for fresh distillation; the dregs however ought to be preserved, as they contain an equal degree of perfume with the essence.

Another method is: To 82 pounds of roses, put 123 pounds of water; distil with a slow heat until about one-half, or rather more if tolerably clear, of the water that comes over. The rose water thus obtained is to be put to another 82 pounds of roses and distilled. The distilled water is then put into broad pans as above. It may be sooner congealed by being put into a large mouthed bottle and subjected to a refrigerating process.

Chapter XX.

List of Patents granted by the United States, for improvements in distillation, on stills, and refining of liquors.

An improvement in distilling. *Aaron Putnam*, January 29th, 1791.

An improvement in distilling. *Nathan Reed*, August 26th, 1791.

Improvement in distilling spirituous liquors. *Joseph Simpson*, March 4th, 1794.

Improvement in a steam still. *Alexander Anderson*, September 2d, 1794.

Improvement on stills. *John Kincaid*, Philadelphia, November 25th, 1794.

Tinned sheet copper condensing worm. *John Taylor*, June 30th, 1795.

Combination of astringent woods and vegetables in distilling. *Fitch Hail*, April 17th, 1797.

Brewing with Indian corn. *A. Anderson*, of Philadelphia.

Condenser for heating wash, in distilling. *A. Anderson*, do. January 26th, 1801.

Improvement in evaporation. *J. Bidwell and Benjamin Henfry*, February 12th, 1801.

Increasing the surface of evaporation, for the purpose of distilling. *Benjamin Henfry*, March 2nd, 1801.

Construction of stills. *Michael Kraft*, Bucks county, Pennsylvania, October 28th, 1801.

Mode of improving spirits. *Burgiss Allison*, New Jersey, May 12th, 1802.

Improvement in a still. *William Paine*, August 24th, 1802.

Do. Do. *John Staples*, December 15th, 1802.

Do. in a boiling cistern. *Timothy Kirk*, December 28th.

Boiler for accelerating the evaporation of liquids.

Improvement in stills. *John Moffat*, February 1st, 1803.

Making brandy out of all kinds of grain, and fruits. *Christ. J. Hutter*, February 11th, 1803.

Improvement in extracting a spirit from starch water. *John Naylor*, March 7th, 1803.

Improvement in a cooler, or condenser of vapour. *William How*, April 6th, 1803.

Improvement in producing steam. *John Stephens*, April 11th, 1803.

Improvement in heating and boiling water. *Benjamin Platt*, April 27th, 1803.

Improvement in distilling spirits, *Daniel Isby*, May 14th, 1803.

Improvement in the application of the principle of rectifying or improving spirits. *Burgiss Allison*, New Jersey, May 17th, 1803.

Improvement in the method of distilling, or making alcohol. *L. J. Kilbourn*, June 4th.

Improvement in distillation, by the application of steam, in wooden, or other stills, or vessels. *Samuel Brown, Edward West and T. West*, July 8th.

Method of. cooling liquors. *David Lownes*, July 21st.

Improvement in the construction of stills. *Lewis Grantz*, August 4th, 1803.

Improvement in boilers, also working stills with the same. *William Thornton*, Washington, October 28th, 1803.

Improvement in still heads and condensers. *Edward Richardson*, December 16th, 1803.

Improvement in stills and boilers. *Leonard Beatty*, January 19th, 1804.

Improvement in the construction of stills, and the process of distilling spirits. *William Wigton*, January 30th.

Improvement in setting stills, and other large kettles. *Israel Wood*, February 21st, 1804.

Improved still and boiler. *John Naylor*, March 31st.

Double steam bath still. *J. I. Giraud*, April 18th.

Method of boiling, applicable to distilling, brewing, &c. *Benjamin Egleton, jun.* Princeton, May 27th, 1805.

Improvement on Anderson's condensing tub. *Henry Witmer*, Lancaster, May 17th, 1805.

Improvement in stills. *Stephen Steward*, June 20th, 1805.

Improvement in condensers in distillation. *Thomas O'Conner*, July 9th, 1805.

Improvement in stills. *Edward Richardson*, October 3d, 1805.

Patent distillery. *Abraham Heistand*, of York county, Pennsylvania, November 27th, 1805.

Machine for raising water. *F. Lippart*, May 17th 1805.

Do. do. *S. Bellig*, November 7th, 1805.

Do. do. *W. Barker*, August 26th, 1808.

Improvement in stills, and rotatory steam engines. *Stedman Adams*, November 21st, 1808.

Improvement in still tubs. *Isaac Bennet*, December 12th, 1808.

Improvement in rectifying spirits. *Osborn Parsons*, June 23d, 1808.

Apparatus for, and mode of distilling. *Abraham Weaver*, Adams county, Pennsylvania, April 23d. 1807.

For distilling or boiling water. *Eli Barnum and Benjamin Brooks*, Berks county, Massachusetts, March 21st, 1808.

A double boiler for distilling. *Amos Gunn*, New Haven, Connecticut, December 19th, 1808.

Improvement in distillation. *John Dimond*, December 28th, 1808.

An apparatus for distilling. *Augustus Tobey*, Berkshire county, Massachusetts, June 24th, 1810.

Improvement in distilling. *Michael Garber, sen.* Staunton, Virginia, July 13th, 1810.

A steam still and water boiler. *P. Bernard*, Whites town, Oneida county, New York.

Perpetual steam still, and water boiler. *P. M. Hackley*, Herkimer county, New York.

Improvement in the water boiler, and steam still. *Borden Wilber and Timothy Soper*, Washington county, New York, February 4th, 1811.

Steam still. *J. N. Walker*, Wilmington, (N. C.) February 9th, 1811.

Improvement in distilling. *Jacob Miller*, Lancaster county, Pennsylvania.

Improvement in distilling. *Andrew Dunlap*, Boston, May 3d, 1811.

Perpetual steam still. *John I. Giraud*, Baltimore, May 15th, 1811.

Improvement in distilling. *J. I. Giraud*, June 11th, 1811.

A distilling and refining apparatus. *Jacob Sherer, and Abraham Killian*, Lancaster.

Improvement in stills, and boilers. *Jacob Decker*, Chenango county, New York, September 2d. 1811.

Improvement in stills. *Henry Witmer*, Lancaster.

A mode of distilling ardent spirits, by one operation. *Jonathan Shaw*, Palmer, Hampshire county, Massachusetts.

A distilling apparatus. *Lewis Lecesne*, New Jersey, December 4, 1811.

A wooden still and condensing box. *Rufus W. Adams*, Vermont, March 27, 1812.

A wooden still head. *Charles Jenks*, Connecticut, April 25, 1812.

In stills. *Hiram Whitecomb*, Connecticut, November 24, 1812.

In distilling. *Omri Carrier*, Connecticut, November 2, 1812.

Do. do. January 18, 1813.

Do. *John Bates*, Connecticut, January 26.

In the still. *J. I. Giraud*, Baltimore, June 26.

In stills. *Charles F. Fisher*, Pennsylvania, July 12.

In the still and condensing tub. *James Wheatley*, Virginia, November 13, 1817.

In water boilers and steam stills. *Timothy Soper and Charles Reynolds*, Connecticut, December 2.

In the distilling apparatus. *Isaac Hoffman*, New York, December 31.

In distilling. *William M. B. Wolias and Hiram P. Barlow*, Pennsylvania, May 6, 1814.

In the still and condenser: *Benjamin Hall*, Connecticut, November 13, 1814.

In distilling. *Napthali Hart*, Pennsylvania, October 29, 1814.

Steam Distillery

A *Boiler*
B *Cylinder*
C *Wash Still*
D *Worm tub*
E *Condensing tub*
F *Rectifying still*

Recommended by A.Anderson & H.Hall

Engraved for Hall's Distiller 2nd Edition

Advertisement.

New Patent Steam Distillery.

The annexed plate represents a plan for a steam distillery, constructed on a new plan, which is superior to any that has hitherto been offered to the public.

The expansive power of steam, and the advantages to be derived from its use as a medium for the conveyance of heat, were long known before any practical application in either way was effected;—at present the arts and sciences receive assistance from this powerful agent.

In the present instance, the subscribers do not make a claim for an original invention, but for a combination of the different methods of using steam, uniting under the same roof an engine for the purpose of grinding all the grain, and doing the principal part of the labour usually done by men, to a boiler for generating steam for the purposes of distilling grain.

The subscribers feel fully satisfied, that the whole machinery necessary for grinding and mashing the grain, and distilling the *wash*, may be erected on the present plan at a much less expense than was paid a few years since for the stills alone.

On the advantages which every distiller would receive from having his own mill, it is quite unnecessary to expatiate. We do not hazard much in the assertion, that for a distillery working 50 bushels a day, the saving in grinding alone in three months would more than pay the cost of the mill machinery.

The boiler and engine necessary for this purpose may be erected at an expense of 1,200 dollars. The cost of the mill machinery and of the distillatory apparatus, can easily be ascertained from the proper mechanics, where the price of the materials is known.

It is not deemed necessary in this publication to enter into a particular description of this plan; the subscribers propose to obtain a patent for it, and will give every satisfaction to any gentleman who may wish to have a distillery erected.

Application may be made by letters (post paid), or to either of us personally.

A. ANDERSON,
HARRISON HALL.

Philadelphia, January, 1818.

CPSIA information can be obtained
at www.ICGtesting.com
Printed in the USA
BVHW082110071221
623481BV00001B/49

9 780991 04368